A MATLAB® Companion for
Multivariable Calculus

A MATLAB® Companion for Multivariable Calculus

Jeffery Cooper
Department of Mathematics
University of Maryland

An Imprint of Elsevier

SAN DIEGO   SAN FRANCISCO   NEW YORK   BOSTON
LONDON   TORONTO   SYDNEY   TOKYO

Senior Acquisitions Editor	Barbara Holland
Production Editor	Vanessa Gerhard
Editorial Coordinator	Karen Frost
Marketing Manager	Marianne Rutter
Cover Design	Dick Hannus
Copyeditor	Elliot Simon
Proofreader	Northwind Editorial Services
Composition	Laser Words
Printer	The Maple-Vail Book Manufacturing Group

MATLAB is a registered trademark of the MathWorks, Inc.

This book is printed on acid-free paper. ∞

ACADEMIC PRESS
An Imprint of Elsevier
525 B Street, Suite 1900, San Diego, CA 92101-4495, USA
http://www.academicpress.com

Academic Press
Harcourt Place, 32 Jamestown Road, London NW1 7BY, UK
http://www.academicpress.com

Harcourt/Academic Press
An Imprint of Elsevier
200 Wheeler Road, Burlington, MA 01803
http://www.harcourt-ap.com

Library of Congress Catalog Card Number: 00-106079
ISBN-13: 978-0-12-187625-8
ISBN-10:   0-12-187625-X

Transferred to Ditital Printing 2009

Contents

Preface

Goals

Multivariable calculus is an essential part of the mathematical education of scientists and engineers. In the past we relied almost entirely on chalk and blackboard, and examples that could be done by hand, to convey the key concepts of the subject. Now the advent of powerful, convenient software makes it imperative to reconsider how we teach multivariable calculus. In my opinion, the greatest impact of software is in the areas of graphics and computation.

Color graphics makes it possible to display curves, surfaces, and solids in two and three dimensions in a way that is both more effective and more engaging for the student. This is especially important for today's students, who have not had much experience with solid geometry. Color can also be used as a fourth dimension to help locate extreme points, and to display quantities such as temperature, density, and fluid velocity.

Software also allows us to deal seriously with computation. Traditionally there has been an unnatural division of analysis and computation into separate compartments of mathematics. The paper-and-pencil exercises of the typical third-semester calculus text give valuable experience in manipulating symbols, but they are often quite contrived. Problems involving arc length, surface, and volume that can be done by hand are too limited in application and give the unfortunate impression that mathematicians are interested only in clever solutions of special problems and prefer to leave practical problems to engineers.

Fortunately, today's software allows us to bridge the gap between analysis and computation. I firmly believe that students must see, and be able to implement, some of the basic numerical techniques. These techniques should not be hidden in a "black box" that is used blindly. Numerical methods depend heavily on linear and quadratic approximation, which are central to calculus. Experience with numerical computations can be used to reinforce the basic concepts of calculus. Furthermore, students (and teachers) should be made aware of the limitations of numerical methods and learn when numerical results are reliable.

This supplement to the traditional third-semester calculus course provides an opportunity to use the graphics capability and the computational power of a complete software package to enrich and broaden the teaching of the subject.

Software

I have chosen to use MATLAB as a software package for several reasons. First, for many of the students taking third-semester calculus, this will be their first exposure to computer work. The easy syntax of MATLAB makes it possible for them to get comfortable with the computer quickly. Second, MATLAB graphics are excellent and easy to use. Third, it is easy to program in MATLAB, making it possible to do numerical calculations using simple loops. One also has a symbolic capability in MATLAB, which uses a part of Maple. Finally, MATLAB is becoming the most popular software package in engineering. Engineering students will see MATLAB in their other courses. Using the same software package in a mathematics course allows for the possibility of joint projects.

I have assumed that the student has access to MATLAB 5.0 or higher, because I introduce inline functions and use some symbolic manipulations. However, one can write an mfile instead of an inline function, and the use of symbolic manipulations is not central to the thrust of the text. Thus, this text can be used satisfactorily with MATLAB 4.x as well.

In this MATLAB companion, I have developed computer-oriented material that complements the usual topics in multivariable calculus. I do not assume that the student knows linear algebra. Vectors and matrices are introduced in the first chapter as arrays of numbers, enough to understand the MATLAB commands. Matrix multiplication is not needed except for some material in Chapters 6, 7, and 8.

Much more material is presented than can be used in one semester. The instructor is free to select the topics most relevant to the course. The main ideas of the standard topics are given without much detail. The emphasis is on examples of the use of MATLAB to display graphs and to do numerical calculations. Many of the examples and exercises are drawn from applications areas such as biology, economics, engineering, and physics. The areas are listed in the following Summary of Contents with the section in which they appear.

The MATLAB examples often contain short bits of code that the student can modify to do the exercises. In addition, I have written a number of MATLAB mfiles to implement some of the graphical displays that I feel would be too time consuming for the beginning student to program. They are available on my Web page at

www.math.umd.edu/~jec

and at the publisher's Web site,

www.harcourt-ap.com/mathematics.html

In some cases, students can write larger codes using these mfiles. For example, there are mfiles that attach arrows in two and three dimensions. These mfiles can be used in the discussion of tangent vectors to curves and surfaces.

There is also a collection of mfiles that can serve as instructor demos or that can be used by the student. These mfiles are discussed in the appendix.

Summary of Contents

A brief introduction to MATLAB is given in Chapters 1 and 2. The material presented there is adequate to get started. Graphing of curves and surfaces in two and three dimensions is introduced as necessary in later chapters.

Chapter 3 is a warmup chapter dealing with lines and planes. Some of the longer exercises require the student to construct graphically structures consisting of lines and planes.

There are a number of sections that present numerical algorithms for solving problems. In Chapter 4, in computing the arc length of a curve, the numerical integrator of MATLAB is used. But in addition, students write their own code to implement the polygonal line approximation. This is needed to compute the arc length of curves given as a table of coordinates, perhaps as the result of a computation. Numerical differentiation is introduced and used. Attention is also paid to the question of estimating the error in these approximations.

In Chapter 5, partial derivatives, directional derivatives, level curves, and tangent planes are illustrated graphically. Some numerical estimates of the error in the tangent plane approximation are made in the exercises.

Graphical tools for displaying level sets of functions of three variables and for displaying parametric surfaces are developed in Chapter 6.

Chapter 7 is devoted to methods of solving systems of equations symbolically and numerically. A two-dimensional Newton's method is presented. An interactive version of Newton's method in two dimensions is included in the set of instructor demos in the appendix.

Chapter 8 has mfiles that permit a student to explore interactively the level curves of a function of two variables and thereby learn to distinguish geometrically between local extreme points and saddle points, as well other behavior. This chapter also has an interactive mfile that can be used to study problems of constrained maxima and minima. Newton's method is used to solve systems of equations that determine the critical points of a function.

In Chapter 9, on multiple integrals, two- and three-dimensional versions of Simpson's rule are implemented and used to compute integrals that cannot be done symbolically but that are commonplace in applications. This gives the student a convenient tool to take away from the course. The formula for change of variable in multiple integrals is illustrated graphically.

In Chapter 10, where surface area is discussed, numerical methods are used to estimate the usual integrals for the surface area. However, in addition, the question of surface area is also addressed using triangular patch approximations, as is done in computer graphics. This approach is then used to find surfaces of minimal surface area, which leads naturally to the problem of minimizing a function of many variables.

Finally, Chapter 11 treats curl, divergence, and the theorems of Green, Gauss, and Stokes. Here mfiles I have written permit the student to explore interactively the circulation of a vector field in two dimensions, thereby illuminating the notion of curl. Similarly, flux integrals are used to give more intuitive content to the divergence.

Chapter 12 deals with problems from electrostatics and fluid flow. The sections on fluid flow are fairly advanced and would certainly be optional material.

Chapter 13 is devoted to features of MATLAB not covered in the brief introduction of Chapters 1 and 2.

Exercises

At the end of each of Chapters 3–12 is a collection of exercises that use MATLAB to illuminate basic concepts of multivariable calculus and to solve application problems.

Application areas are treated in the exercises as follows:

Engineering: 3.4, 3.7, 3.11, 3.12, 3.13, 4.11, 4.12, 6.8, 6.15, 6.17, 7.8, 8.6, 9.6, 10.5, 10.6, 10.7
Economics: 8.15
Physics: 4.5, 4.10, 5.9, 6.3, 6.4, ·8.10, 8.14, 9.13, 10.8
Biology: 5.10, 8.16
Animal science: 9.14

Many of the exercises are quite short and can be done in a few lines on the command line. Nevertheless, the graphs do need some interpretation. Exercises marked with ⋆ require several steps and usually a short mfile. Longer problems, marked with ⋆⋆, can serve as computational projects.

Because I believe the student should become comfortable with MATLAB as a tool to be used easily and frequently, in my own classes I usually assign several shorter exercises every week and a longer problem every two or three weeks.

If possible, one or two classes during the first week should be conducted in a computer lab where students can run through the material of the first two chapters under the guidance of the instructor.

Advanced Matlab

Many books have been written about MATLAB in addition to the manuals prepared by Mathworks (publishers of MATLAB). They come in various shapes and forms, from elementary introductions to advanced treatises on MATLAB graphics, which is a whole subject in itself. Here are two more advanced texts.

MATLAB Guide by Desmond J. and Nicholas J. Higham, SIAM (Society for Industrial and Applied Mathematics), 2000.
Introduction to Scientific Computing, A Matrix-Vector Approach using MATLAB, 2nd ed., Charles F. Van Loan, Prentice Hall, 1999.

Web page

Additional exercises and programs as they are developed will be posted on the author's Web page: www.math.umd.edu/~jec. Readers are encouraged to send comments and suggestions to the author at jec@math.umd.edu.

Acknowledgements

I wish to thank my colleagues at the University of Maryland for their many suggestions and helpful comments during the preparation of the manuscript. In particular, I have had many stimulating conversations with John Osborn on the subject of the computer in mathematics courses. The reviewers of earlier versions of the text made many suggestions that improved the final product. Kevin Gormally and Andrey Rukhin made valuable contributions by working through many of the exercises. Finally, I wish to thank the many students who used these materials and made valuable suggestions.

List of mfiles

We list here the prepared mfiles, according to the chapter in which they are introduced. These files are available on my Web page at

www.math.umd.edu/~jec

Chapter 3

    arrow.m arrow3.m plane.m

Chapter 4

    frenet.m

Chapter 5

    xslice.m yslice.m mslice.m qsurf.m

Chapter 6

    impl.m

Chapter 7

    newton2.m

Chapter 8

    findcrit.m lagrange.m

Chapter 9

    riemann.m simp2.m simp3.m trf.m

Chapter 10

    tsurf.m gdome.m

Chapter 11

    lint.m flux2.m curl.m

Chapter 12

    flow1.m flow2.m

Appendix

    findroot.m minsurf.m flux3.m circ.m tplane.m tarea.m

1

Basic MATLAB: The Command Line

In this chapter, we discuss operations that can be performed from the command line. In Chapter 2, we discuss mfiles and programs.

1.1 First steps

When you invoke MATLAB to begin a session, you see the prompt

```
>>
```

When we give instructions for an operation, or request information, following this prompt, we say that we are working on the *command line*.

We can do simple arithmetic operations on the command line, such as $(2 + 3.5^2 - 4 \cdot 7)/12$,

```
>> (2+3.5^2 -4*7)/12
ans =
  -1.1458
```

We can also do this calculation by assigning variable names to the quantities:

```
>> x = 2+3.5^2
ans =
  14.2500
>> y = 4*7
ans =
  28
>> z = (x-y)/12
ans =
  -1.1458
```

If we do not wish to see the intermediate results, we can suppress the numerical output by putting a semicolon at the end of the line. Then the sequence of commands and output looks like this:

```
>> x = 2+3.5^2;
>> y = 4*7;
>> z = (x-y)/12;
>> z
z =
    -1.1458
```

MATLAB does numerical calculations in double precision, which is 15 digits. Normally only five digits are displayed. If we want to see all 15 digits, we use the command `format long`:

```
>> format long
>> z
z =
    -1.14583333333333
```

To return to the short format, enter `format short`.

Error messages

If we enter an expression incorrectly, MATLAB will return an error message, which sometimes locates the error. For example, in the following, we left out the `*` in `3*x`:

```
>> x = 4;
>> 3x
??? 3
     |
Missing operator, comma, or semicolon.
```

Another example.

```
>> 2*(x+y
??? 2*(x+y
         |
A closing right parenthesis is missing.
Check for a missing ")" or a missing operator.
```

Making corrections

To make corrections, we can, of course, retype the expression. But if the expression is lengthy, we may make more mistakes by typing a second time. Unfortunately we cannot move the cursor to the line we wish to repair. Instead we can press the up-arrow key until we reach the desired line and then the left- and right-arrows until we reach the offending characters. Type in the correction and enter return.

Exiting

To leave MATLAB enter `quit`.

    If MATLAB gets hung up in calculation or is taking a long time, and you want to stop the calculation, without exiting MATLAB, enter Ctrl+C.

HELP ! !

Help with most operations is available with a keystroke, thanks to the online help provided by MATLAB. To get information on a particular command or operation, simply enter `help` *command name*. For example, to get information on how to use the plotting commands, enter `help plot`.

1.2   Vectors and matrices

Vectors and matrices are the basic elements of the MATLAB environment. In this text we shall be using the word *vector* in two, related ways.

    In Chapter 3, we shall speak of vectors as directed line segments in two- and three-dimensional space, used to represent physical and geometric quantities such as force and velocity.

    In this chapter, we shall use vector to mean an ordered list of numbers, written either horizontally or vertically. For example,

$$\mathbf{u} = [2, 1.3, \sqrt{2}, 8, -4, \pi]$$

or

$$\mathbf{v} = \begin{bmatrix} 1 \\ -2 \\ \pi \\ 4.2 \end{bmatrix}.$$

We say that $\mathbf{u}$ is a row vector and that $\mathbf{v}$ is a column vector.

A *matrix* is a rectangular array of numbers. For example,

$$\mathbf{A} = \begin{bmatrix} 1 & 2 & 3 & 9 \\ 4 & 5 & 6.1 & -2 \\ \pi/2 & 1/3 & 4 & -1 \end{bmatrix}.$$

The dimensions of a matrix are the number of rows and the number of columns, with the number of rows usually given first. The matrix $\mathbf{A}$ here is a 3×4 matrix. The row vector $\mathbf{u}$ is a 1×6 matrix, and the column vector $\mathbf{v}$ is a 4×1 matrix. A single number, such as 5.2, is a *scalar* and can be considered a 1×1 matrix. The entries in a matrix often are written $a_{i,j}$, with i being the row index and j being the column index. For example, in our matrix $\mathbf{A}$, $a_{2,1} = 4$ and $a_{3,2} = 1/3$.

The *transpose* of an $m \times n$ real matrix $\mathbf{A}$ is the $n \times m$ matrix that results from interchanging the rows and columns of $\mathbf{A}$. The transpose matrix is denoted $\mathbf{A}^T$. The transpose of our matrix $\mathbf{A}$ is

$$\mathbf{A}^T = \begin{bmatrix} 1 & 4 & \pi/2 \\ 2 & 5 & 1/3 \\ 3 & 6.1 & 4 \\ 9 & -2 & -1 \end{bmatrix}.$$

Various operations can be performed on vectors and matrices and we shall illustrate them in the context of MATLAB.

Forming vectors and matrices

Matrices can be entered by typing in the elements one at a time. To enter the matrix

$$\mathbf{A} = \begin{bmatrix} 1 & 2 & 3 \\ 4 & 5 & 6 \end{bmatrix},$$

we type

```
>> A = [1 2 3;4 5 6]

A =
     1     2     3
     4     5     6
```

Notice that we use a semicolon to separate the rows. Remember, to suppress the output, put a semicolon after the defining statement. This can be especially important if the matrix or vector has thousands of elements.

The transpose of a real matrix is formed by the command A'. If the row vector x is defined by

```
>> x = [1 5 4 8 10]
```

then x is turned into a column vector with the command x'. If the matrix or vector has complex elements, the command A' produces the Hermitian transpose, which is the transpose with the complex conjugate of the elements. For example,

```
Z =
     1+i    2    1
     2+5i   i    2
>> Z'
Z'=
     1-i   2-5i
      2     -i
      1      2
```

To get a transpose without taking the complex conjugates, use A.'. Note that we put a dot before the apostrophe.

To determine the dimensions of a vector or matrix, use the command size, as follows:

```
>> size(A)
ans =
      2       3

>> size(x)
ans =
      1       5

>> size(x')
      5       1
```

We can view a particular element in a vector or matrix by specifying its location:

```
>> A(1,2)
ans =
      2

>> x(5)
ans =
     10
```

Often we must deal with vectors or matrices that are too large to enter one element at a time. If there is some formula or some regular pattern to the elements, we may be able to use special commands. For example, suppose we want to enter a vector **x** consisting of points $(0, .1, .2, .3, .4, \ldots, 5.9, 6)$. We can use the command

```
>> x = 0:.1:6 ;
```

This row vector has 61 elements. Another way to create the same vector is to use the command `linspace` (which stands for "linear spacing"):

```
>> x = linspace(0,6,61);
```

This is useful when we want to divide an interval into a number of subintervals of the same length. For example, `theta = linspace(0, 2*pi, 41)` divides the interval $[0, 2\pi]$ into 40 equal subintervals, creating a vector of 41 elements.

To create a vector of zeros or of ones of the same dimensions as a given vector x, there are commands

```
>> y = ones(size(x));
>> z = zeros(size(x));
```

The same works for matrices

```
>> Z = zeros(size(A));
>> Y = ones(size(A))
Y =
        1      1      1
        1      1      1
```

We can also specify a matrix of zeros or ones by giving the dimensions:

```
>> Z = zeros(2,3)
```

The $n \times n$ identity matrix is produced with the command `eye(n)`. There are special commands for entering sparse matrices or diagonal matrices. For more information, enter `help sparse` or `help diag`.

1.3   Array operations

Arithmetic of matrices

There is an obvious, natural way to add and subtract matrices:

```
>> B = [2  0  -1;  1  2  7];
>> A + B
ans =
        3      2      2
        5      7     13
```

Usually, we can add together only matrices having the same dimension. There is an exception in MATLAB, however, that is very useful. Suppose we want to add the same number c to each element of a matrix $\mathbf{A}$. This can be done with the command A + c*ones(size(A)), or more simply, $A + c$. In particular, if $\mathbf{x}$ is a vector, we can add a scalar t to each component of $\mathbf{x}$ with the command x+t.

We can always multiply a matrix by a scalar or divide by a nonzero scalar.

```
>> 2 * A
ans =
        2      4      6
        8     10     12

>> A/2
ans =
   0.5000    1.0000    1.5000
   2.0000    2.5000    3.0000
```

Array operations

Arithmetic operations can also be performed on matrices, entry by entry. These are called *array operations*. Array multiplication is an example. If $\mathbf{A}$ and $\mathbf{B}$ are two matrices of the same size with elements $a_{i,j}$ and $b_{i,j}$, then the command

```
>> C = A.*B
```

produces another matrix $\mathbf{C}$ of the same size with elements $c_{i,j} = a_{i,j}b_{i,j}$. For example, using the same 2×3 matrices $\mathbf{A}$ and $\mathbf{B}$ we defined earlier, we have

```
>> C = A.*B
C =
        2      0     -3
        4     10     42
```

To raise a scalar to a power, say, 2, we use the command 5^2. If we want the operation to be applied to each element of a matrix, we use .^2. For example, if we want to produce a new matrix whose elements are the square of the elements of the matrix $\mathbf{A}$ we enter

```
>> A.^2
ans =
       1      4      9
      16     25     36
```

There is also a kind of array division for two matrices of the same size that divides the two matrices element by element:

```
>> D = [1 3 5; -2 4 -1]
>> A./D
ans =
       1.0000      0.6667      0.6000
      -2.0000      1.2500     -6.0000
```

1.4 Matrix multiplication and linear systems

Another kind of multiplication between matrices is motivated by the consideration of linear systems of equations. Let $\mathbf{A}$ be the 2×3 matrix

$$\mathbf{A} = \left[\begin{array}{ccc} a_{1,1} & a_{1,2} & a_{1,3} \\ a_{2,1} & a_{2,2} & a_{2,3} \end{array} \right]$$

and

$$\mathbf{x} = \left[\begin{array}{c} x_1 \\ x_2 \\ x_3 \end{array} \right],$$

a 3×1 column vector. We define the product $\mathbf{Ax}$ to be a 2×1 column vector with components

$$\left[\begin{array}{c} a_{1,1}x_1 + a_{1,2}x_2 + a_{1,3}x_3 \\ a_{2,1}x_1 + a_{2,2}x_2 + a_{2,3}x_3 \end{array} \right].$$

With this definition of multiplication of a matrix by a vector, we can write the linear system of two equations in the three unknowns x_1, x_2, x_3,

$$a_{1,1}x_1 + a_{1,2}x_2 + a_{1,3}x_3 = b_1$$
$$a_{2,1}x_1 + a_{2,2}x_2 + a_{2,3}x_3 = b_2,$$

as simply

$$\mathbf{Ax} = \mathbf{b},$$

where $\mathbf{b}$ is the 2×1 column vector

$$\left[\begin{array}{c} b_1 \\ b_2 \end{array} \right].$$

More generally, if $\mathbf{A} = [a_{i,j}]$ is an $m \times n$ matrix and $\mathbf{x} = [x_1, x_2, \ldots, x_n]$ is an $n \times 1$ column vector, we define $\mathbf{Ax}$ to be the $m \times 1$ column vector with ith component

$$\sum_{j=1}^{n} a_{i,j} x_j.$$

In this way, the system of m linear equations in n unknowns x_j,

$$\sum_{j=1}^{n} a_{i,j} x_j = b_i, \qquad i = 1, \ldots, m$$

can be written compactly as

$$\mathbf{Ax} = \mathbf{b}. \tag{1.1}$$

Now let $\mathbf{A}$ be an $m \times n$ matrix and $\mathbf{B}$ be an $n \times p$ matrix. We label the columns of $\mathbf{B}$ as $\mathbf{B}_j = [b_{i,j}]$, $j = 1, \ldots, p$. We define

$$\mathbf{AB} = \mathbf{C}, \tag{1.2}$$

where $\mathbf{C}$ is the $m \times p$ matrix whose columns are the $m \times 1$ column vectors $\mathbf{C}_j = \mathbf{AB}_j$, $j = 1, \ldots, p$. In terms of the entries,

$$c_{i,j} = \sum_{k=1}^{n} a_{i,k} b_{k,j}.$$

This matrix multiplication $\mathbf{AB}$ is only defined for an $m \times n$ matrix $\mathbf{A}$ and an $n \times p$ matrix $\mathbf{B}$. The column dimension of $\mathbf{A}$ must equal the row dimension of $\mathbf{B}$.

In MATLAB we can multiply matrices in this fashion with the * symbol. It is very important to notice that this kind of matrix operation uses the symbol *, without the dot in front. Remember, we use the symbol .* for array multiplication. We assume we have matrices of the correct dimensions:

```
>> A = [1 2; 3 3; 4 5];
>> B = [-1 3; 5 1];
>> C = A*B;
>> C
=   9    5
   12   12
   21   17
```

If $\mathbf{A}$ is a square matrix, $n \times n$, $\mathbf{A}$ can be multiplied times itself any number of times. We use the notation $\mathbf{A}^k$ to denote the product of k factors $\mathbf{AA}\ldots\mathbf{A}$. The MATLAB command for raising a matrix to a power is A^k. Notice that the command does *not* have the dot in front. A.^k means the array operation that raises each element of $\mathbf{A}$ to the kth power.

Given an $n \times n$ matrix $\mathbf{A}$ and an n column vector $\mathbf{b}$, the linear system $\mathbf{Ax} = \mathbf{b}$ can be solved in several ways. The simplest way is to use the command A\b:

```
>> A = [1 2 3; 4 5 6; 6 7 9];
>> b = [ 1 0 1]';
>> x = A\b;
x =
    -0.0000
    -2.0000
     1.6667
```

The command A\b uses the method of Gaussian elimination with partial pivoting to solve linear systems.

1.5   MATLAB functions

MATLAB basically has two kinds of functions, numerical functions and symbolic expressions of functions. A *numerical function* is really a short program that operates on numbers to produce numbers. A *symbolic expression of a function* operates on symbolic variables to produce symbolic results. These symbolic expressions can be manipulated with operations such as differentiation and integration. We shall discuss symbolic expressions of functions in the next section.

MATLAB has the usual built-in numerical functions, such as $\sin x$, $\cos x$, $\tan x$, $\exp x$, $\log x$, and $\sqrt{x}$. These functions can take matrices as arguments, in which case the function is applied to each element of the matrix. We say that such a function is *array-smart*. For example, the cosine function can be applied to a matrix:

```
>> T = [2 3 pi; 8 pi/2 1];
>> cos(T)
ans =
    -0.4161    -0.9900    -1.0000
    -0.1455     0.0000     0.5403

>> sqrt(A)
ans =
     1.0000     1.4142     1.7321
     2.0000     2.2361     2.4495
```

In addition, many other specialized functions are available. These include the error function, called with the command erf(x), and Bessel functions of all orders. There are also functions of linear algebra that find information about matrices, such as eig(A), which finds the eigenvalues of a matrix **A**.

Nevertheless, we will often need to build our own numerical functions of one, two, or three variables. In this section we shall consider only functions of one variable. Functions of several variables will be discussed in a later chapter.

Prior to version 5.0 of MATLAB, numerical functions could be constructed only in separate files called *mfiles*. This way of constructing functions will be covered in Chapter 2.

Now with versions 5.0 and higher, there is an easy way of constructing a numerical function on the command line. This kind of numerical function is called an *inline function*. Here is a simple example:

```
>> f = inline('x^3 +x -1')
```

To evaluate $f(x) = x^3 + x - 1$ at $x = 2$, enter f(2). If we wish the function to be array-smart, we must write

```
>> f = inline('x.^3 +x -1')
```

Here we have used instruction .^ for the array operation. Functions created this way can accept vectors and matrices as arguments. The function will be applied to each element of the vector or matrix. For example, if the matrix **A** is given by

```
A =
     1     2     3
     4     5     6
```

then

```
>> B = f(A)

B =    1     9    29
      67   129   221
```

We shall need our numerical functions to be array-smart to do many computations and for the purposes of graphing.

One of the most common mistakes of beginners is to forget to make their numerical functions array-smart by inserting the dot before the operations *, /, and ^.

Unfortunately, we cannot add or multiply inline functions to produce a new function. If we define the inline function g by the command

```
>> g = inline('cos(x) + x')
```

we *cannot* use the command

```
>> h = f+g
```

to produce the function $f + g$. Instead we must define a new inline function

```
>> h = inline('x.^3 + 2*x - 1 + cos(x)')
```

1.6 Symbolic calculations

Up to this point, we have been using MATLAB on your computer as a large, sophisticated calculator. For example, if we enter a matrix **A** of numbers, we can find its determinant as a number. We have also created numerical functions. However, MATLAB also has the capability to manipulate expressions symbolically. There are tools to perform algebraic operations, differentiate and integrate functions, solve systems of equations, and solve ordinary differential equations. These tools come from the software program Maple developed at the University of Waterloo, Canada.

Creating symbolic expressions

Variables x, y, z, a, b, c, etc. can be declared symbolic variables with the command

```
>> syms x y z a b c
```

This command is a shortcut for the more elaborate command `sym('x', 'y', 'z', 'a', 'b', 'c')`, or even more deliberately, `x = sym('x')`, `y = sym('y'), . . . .` We can then define expressions using these variables, and these expressions can be manipulated symbolically. For example, a matrix **A** can be defined by

```
>> A = [ a b 1; 0 1 c; x 0 0 ]
A =
[a, b, 1]
[0, 1, c]
[x, 0, 0]
```

Since `A` is a symbolic expression, we can calculate its determinant in terms of the variables a, b, c, x with the usual MATLAB command:

```
>> d = det(A)
d = x*(b*c-1)
```

Functions defined symbolically

A function $f(x)$ can be defined in terms of a symbolic expression by this kind of command.

```
>> f = a*x^2 + b*x +c + 2*cos(x)
```

Notice that we do not use the array operations .^, .*, ./ in symbolic expressions, because symbolic expressions are not applied directly to vectors and matrices.

The symbolic expression for this function cannot be evaluated with the simple command f(2). We will need another set of commands, which are explained a bit further on.

Now we can differentiate this symbolic expression with the command (and output)

```
>> diff(f)
ans = 2*a*x+b-2*sin(x)
```

MATLAB differentiates with respect to the variable closest to x in the alphabet. If we wish to differentiate f with respect to the variable a, we must specify that in the command: diff(f,a). If we wish to make further operations on the derivative, we can give it a name, which will be the name for another symbolic expression:

```
>> fprime = diff(f)
fprime = 2*a*x+b-2*sin(x)
```

The second derivative can be computed by differentiating the expression fprime or by using a variation on the diff operation,

```
>> diff(f,2)
>> 2*a-2*cos(x)
```

Higher derivatives are calculated with diff(f,3), diff(f,4), etc.

We can also find the antiderivative of functions defined symbolically. For example, using the same function just defined, we have

```
>> int(f)
ans =
1/3*a*x^3 + 1/2*b*x^2 +c*x +2*sin(x)
```

This operation provides us with an indefinite integral, to which we may add any constant. To compute the definite integral, over, say, [0, 3], we use the command

```
>> int(f,0,3)
ans =
9*a +9/2*b +3*c+2*sin(3)
```

Here we assumed that we wanted to integrate the expression with respect to the variable x. If instead, we wanted to consider a as the variable of integration, we must specify that, with the command

```
int(f,a)
ans =
1/2*a^2*x^2 +b*x*a +c*a+2*cos(x)*a
```

Many other variations are possible. To see them, enter `help sym/int.m`.

Evaluating symbolic expressions

Next, how do we specify the values of the parameters in the expression, and how do we evaluate the symbolically defined function at a point? This is done using the substitution command, `subs`. The syntax is `subs(f,old,new)`, where the old values of the parameters and variables are replaced by new values.

For example, if we wish to evaluate the function f defined earlier at $x = 2$, leaving in the parameters a, b, c, we enter

```
>> subs(f,x,2)
ans =
9*a+3*b+c+2*cos(3)
```

The result is still a symbolic expression. If we wish to specify the values of the parameters, say, $a = 2$, $b = -3$, $c = 9$, we do it this way:

```
>> g = subs(f, [a b c], [2 -3 9])
g =
2*x^2-3*x+9+2*cos(x)
```

Now we have a symbolic expression depending on the one variable x. To evaluate this function at a particular point, say, $x = -1.5$, we can make another substitution, `subs(g,x,-1.5)`, with the answer of `18 + 2*cos(3/2)`. The result is still a symbolic quantity. If we wish to convert it to a floating point number in double precision, we use `double(18 + 2*cos(3/2))`, or in one command as `double(subs(g,x,-1.5))`. Again, many variations are possible. For further information, enter `help sym/subs.m`.

In Chapter 13, we discuss how to convert symbolic expressions to inline functions. This is important for graphing functions of several variables that arise in symbolic computations.

Solving equations symbolically

MATLAB can also solve certain equations symbolically, in terms of parameters in the equation. For example, to solve the equation $ax^2 + bx + c = 0$ we define the symbolic variables x, a, b, c and the expression $f = ax^2 + bx + c$ with commands

```
>> syms x a b c
>> f = a*x^2+b*x+c
>> solve(f)
ans =
[ 1/2/a*(-b+(b^2-4*a*c)^(1/2))]
[ 1/2/a*(-b-(b^2-4*a*c)^(1/2))]
```

Of course, these are the two solutions of the quadratic formula. The command `solve` assumes you want to solve the equation $f(x) = 0$.

For another example, consider the equation

$$\ln(y) - \ln(r - y) = kt + C.$$

To solve for y in terms of t, r, k, and C, we can use the symbolic expression for $f = \ln(y) - \ln(r - y) - kt - C$. The command `solve(f,y)` tells MATLAB to solve the equation

$$f(t, y, r, k, C) = 0$$

for y in terms of the other variables:

```
>> syms t y r k C
>> f = log(y) - log(r-y) - k*t - C
>> y = solve(f,y)
y =
r/(1+exp(k*t+C))*exp(k*t+C)
```

We can then find that value of t such that $y = 5$, in terms of the other parameters r, k, C, with the commands

```
>> solve(y-5,t)
ans =
-(-log(5/(r-5))+C)/k
```

We shall investigate how to solve systems of equations involving several variables in later chapters.

1.7   Two-dimensional graphs

Graphing numerical functions

MATLAB has an excellent set of graphic tools. In this section we will touch on only some of the most elementary ones. We begin with two-dimensional graphs. The basic MATLAB graphing procedure in two dimensions is to take a vector of x coordinates, $\mathbf{x} = (x_1, \ldots, x_N)$, and a vector of y coordinates, $\mathbf{y} = (y_1, \ldots, y_N)$, locate the points (x_j, y_j), and then join them by straight lines. The command is plot(x,y). The vectors $\mathbf{x} = (1, 2, 3, 4, 5)$ and $\mathbf{y} = (-1, 2, 3, 1, 5)$ plotted this way produce the picture shown in Figure 1.1.

```
>> x = [1 2 3 4 5];
>> y = [-1 2 3 1 5];
>> plot(x,y)
```

We graph a numerical function in the same way. For example, to graph the function $\cos x$ on the interval $[-\pi, \pi]$, we first create a vector of x coordinates. Then we create a vector of y coordinates that are the values of $\cos x$ at these points. Finally, the points are plotted and joined by straight lines:

```
>> x = linspace(-pi, pi, 51)
>> y = cos(x);
>> plot(x,y)
```

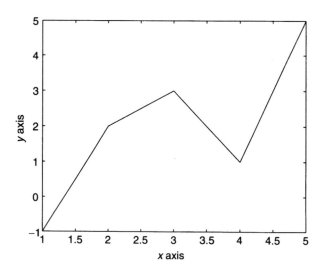

Figure 1.1 Plot **x** versus **y** for the vectors $\mathbf{x} = (1, 2, 3, 4, 5)$ and $\mathbf{y} = (-1, 2, 3, 1, 5)$.

If the function f is defined as an inline function, we can graph it with the command plot(x,f(x)). For example, if we want to plot $f(x) = x^3 + x - 1$ on the interval $[1, 5]$, we use the commands

```
>> f = inline('x.^3 + x -1')
>> x = linspace(0,5, 101);
>> plot(x, f(x))
```

The color of a single curve in MATLAB 5.0 or higher is, by default, blue, but other colors are possible. The desired color is indicated by a third argument, which is a character string. For example, red is selected by plot(x,y,'r'). Note the single quotes around r. The color table is

y	yellow
m	magenta
c	cyan
r	red
g	green
b	blue
w	white
k	black

For a complete listing of the combinations of colors and symbols, enter help plot.

There are two ways we can plot several curves on the same graph. Remember, a curve is determined by a pair of vectors $\mathbf{x}$, $\mathbf{y}$, each with the same dimensions $n \times 1$ or $1 \times n$. Suppose there is another pair of vectors, $\mathbf{z}$, $\mathbf{w}$, with dimensions $m \times 1$ or $1 \times m$, where m may differ from n. The first way to plot the two curves on the same graph is with the command

```
>> plot(x,y,z,w)
```

In MATLAB 4.2 the first curve will be in yellow, the second in magenta. In MATLAB 5.0 and higher, the colors will be blue and green.

Two functions f and g given as array-smart inline functions can be plotted on $[-1, 4]$ together with $\exp(x)$ by the commands

```
>> x = -1:.1:4;
>> plot(x,f(x),x,g(x),x,exp(x))
```

The three curves will be in different colors.

The second way to plot several curves on the same graph uses the command hold on:

```
>> plot(x,y)
>> hold on
>> plot(z,w)
>> hold off
```

Both curves will now be the same color. The three functions $f(x)$, $g(x)$, and e^x are plotted together with these commands:

```
>> plot(x,f(x))
>> hold on
>> plot(x,g(x))
>> plot(x,exp(x))
>> hold off
```

The ezplot command

The command `ezplot` is used primarily to graph functions that are defined symbolically. If f is defined by a symbolic expression and we wish to graph it on the interval $[1, 5]$, we can do it with the one command, `ezplot(f, [1,5])`. For example,

```
>> syms x
>> f =  cos(x)^2*exp(x)
>> ezplot(f, [1,5])
```

This can be most useful after a symbolic calculation leads to a complicated expression. Using the function $f(x) = (\cos x)^2 \exp x$, if we want to quickly graph the second derivative of f, we could add the lines

```
>> g = diff(f,2)
>> ezplot(g, [1,5])
```

The `ezplot` command picks its own points for graphing, using more where the function changes rapidly and fewer where it changes more slowly.

In versions 5.2 and higher of MATLAB, the `ezplot` feature has been extended to graph curves given parametrically in two or three dimensions (with animation). It has also been extended to graph functions of two variables. We shall see these new features as each topic is considered.

Further graphing features

Labels and a title can be attached to the graph with additional commands, for example,

```
>> xlabel(' t, time after lift off,
                                in seconds ')
>> ylabel(' h, height above ground in meters    ')
>> title(' vertical climb of rocket    ')
```

The axis *command.* When we use the command plot(x,y), MATLAB automatically plots the curve on the rectangle $[x_{min}, x_{max}] \times [y_{min}, y_{max}]$. If we wish to change this scale, perhaps to expand a portion of the graph, and instead plot on the rectangle $[a, b] \times [c, d]$, we follow the plot command with axis([a b c d]). You can return the axis scaling to the automatic (default) mode with the command axis('auto') (alternate form axis auto).

The zoom *command.* This is another way to enlarge a portion of the graph, using the mouse. Enter the command zoom on. Then move the pointer to the region of the graph you want to blow up. Click with the left mouse button. This will enlarge the portion by a factor of 2. Clicking again enlarges it again by a factor of 2. Clicking with the right mouse button has the opposite effect. The command zoom out restores the original figure. zoom off turns off the zoom feature.

1.8 Managing the workspace and getting help

Now that you can solve some equations and graph some functions, you will find the following utility commands very useful.

Workspace commands

These commands allow you to find what you have in your workspace and how to clear out unneeded variables.

who lists variables currently in the workspace and their type.

clear clears the workspace; all variables are removed.

clear x y g removes only the variables x, y and the function (either inline or symbolic) g.

clf clears the figure window.

close closes the figure window.

Getting information

Remember if you know the name of the command or feature and want information about it, enter help *command name*. If a command calls an mfile, and you want to see the code of that mfile displayed on the screen, enter type *command name*. For example, the MATLAB feature fzero finds the zeros of a function of one variable. For information on how to use it, we enter help fzero. To see the code, we enter type fzero. To find where in the structure of directories fzero can be found, we enter which fzero.

Some of the help files and codes are rather long, and they go by on the screen very quickly. To see them one screen at a time, enter more on before entering any of the query commands. When you are done, enter more off.

All this information is very accessible if you know the name of the command. However, suppose you want to know if MATLAB has a command, or several commands, that deal with a certain kind of problem. In this case we use the command lookfor. For example, suppose we want to know if MATLAB has a function that finds the largest element of a vector or matrix. We would enter lookfor largest. The professional version yields the following listing:

```
>> lookfor largest

REALMAX Largest positive floating point number.
MAX     Largest component.
NNFMC Find largest column vector in matrix.
```

When we enter help max we find

```
>> help max

MAX     Largest component.
    For vectors, MAX(X) is the largest element in X. For
    matrices, MAX(X) is a row vector containing the maximum
    element from each column. For N-D arrays, MAX(X) operates
    along the first non-singleton dimension.
                            .
                            .
                            .

    See also MIN, MEDIAN, MEAN, SORT.
```

2

Basic MATLAB: mfiles

We discuss how to create and edit files in MATLAB. This is followed by a description of function mfiles and script mfiles. We finish the chapter with instructions on how to save work, print out figures, and prepare documents.

2.1 Creating and editing files in MATLAB

Working in MATLAB from the command line is virtually independent of the type of machine you are using. The different versions of MATLAB for PC, Mac, and Unix machines are adapted to run the same way on each of these platforms. However, when we venture beyond the command line, there are differences. We shall need to create and edit files, called *mfiles*, to

1. Create and save more complicated functions
2. Write and record longer sequences of commands

PCs and Macs

On PCs and Macs, MATLAB provides its own editor. In the upper left corner of the command window, click on the word "File." This opens the "File" menu. To write a new mfile, click on the line "New." This will bring up another window, the MATLAB Editor/Debugger. After writing your file, usually a sequence of MATLAB commands, open the "File" menu of the Editor/Debugger window. Then you can name your file and save it with the "Save as" command. Usually, this will save your file in the current working directory and MATLAB will be able to find it when you call for it from the command line. However, if you are working on a shared system, there may be different arrangements and you must check with the system manager.

After you have saved your file, do not close the Editor/Debugger window. All too often, there is an error in the sequence of commands and you must return to the file to change it. With the file still in the Editor/Debugger window, you can make changes. However, these changes will not be recorded until you again go to the "File" menu of the Editor/Debugger window and click on "Save."

Unix machines

Prior to version 5.2, Unix versions of MATLAB did not provide their own editor. In this case you must use your choice of Unix editor, such as `vi`, `emacs`, or `pico`, in a separate window into the same working directory. It is possible to work from the MATLAB command window by entering the command `!vi [file name]`. In this case, MATLAB turns over control to the local system until you have finished editing the file.

With versions 5.2 and higher of MATLAB the command `edit` brings up the Editor/Debugger window, and you can use it as if you were working on a PC.

Now you may be impatiently asking, what kind of files will we be writing?

2.2   Mfiles

Mfiles are a very convenient, flexible way of collecting sequences of commands that may be lengthy or tedious to type over and over again. Mfiles may be saved to be used at another time. There are two kinds of mfiles: function mfiles and script mfiles. The names of mfiles always have the extension `.m`.

Function mfiles

Function mfiles are used mostly to write numerical functions whose expression is long or complicated and that we want to save for future use.

Suppose we need to compute the values of the function

$$f(x) = x \exp(-\sin x)/(1 + x^2).$$

We can create a function mfile, called `f.m`, so that to evaluate f at $x = 2$, we need only enter `f(2)` on the command line. The mfile is a file that should be placed in the same directory where you are using MATLAB. Here is what the mfile looks like. Function mfiles always begin with a function statement.

```
function y = f(x)
y = x*exp(-sin(x))/(1+x^2);
```

Written this way, the function can take only scalars for x. However, if we write it using the symbols for the array operations, like this,

```
function y = f(x)
y = x.*exp(-sin(x))./(1+x.^2);
```

the function is now array-smart and can be used on vectors and matrices. Notice, in the denominator we are adding the scalar 1 to the vector x.^2 to produce another vector, which then divides in array fashion the factor x.*exp(-sin(x)).

Functions that are defined piecewise may also be constructed in an array-smart fashion. Consider the example

$$
f(x) = \begin{cases} x & x < 0 \\ x^2 & 0 \le x < 2 \\ 4 & x \ge 2 \end{cases}.
$$

The building blocks for this kind of function are the *characteristic* functions for intervals of the form $(-\infty, a)$ and (a, ∞). For example, the characteristic function for $(-\infty, a)$ is $c(x) = 1$ for $x < a$ and $c(x) = 0$ for $x \ge a$. We use the MATLAB logical expression $(x < a)$. When applied to a scalar x, this function returns a 1 if the inequality is true and a 0 if it is false. When applied to an n vector $x = (x_1, \ldots, x_n)$, the logical function $(x < a)$ returns an n vector of 0's and 1's, with a 1 whenever the inequality is true and a 0 whenever it is false. The logical functions $(x > a)$ and $(x <= a)$ work in the same way. An mfile for the characteristic function of the interval $(-\infty, 3)$ would be

```
function y = c(x)
y = (x < 3);
```

Check that $c(x) = 1$ for $x < 3$ and $c(x) = 0$ for $x \ge 3$. Now we make an mfile for f that is array-smart as follows:

```
function y = f(x)
y1 = x.*(x < 0);
y2 = x.^2.*( (x < 2) - (x < 0) );
y3 = 4*(1 - (x < 2));
y = y1 + y2 + y3;
```

Finally, we note that the variables used in the mfile to define the function are "dummy" variables. We can use any variable names to call the function. For example, for the function f defined here, we can use the statements

```
s = -2:.1:4;
r = f(s);
```

The first command defines the vector s with 61 components, and the second command computes another vector r with $r_i = f(s_i)$ for $i = 1, \ldots, 61$.

Summary of function construction

We have now seen three ways to create functions with MATLAB.

Numerical functions are constructed using inline functions (Section 1.5) and function mfiles (this section).

Symbolic expressions for function are constructed, and manipulated, using the symbolic operations described in Section 1.6.

Graphing

The command `plot` works with numerical functions defined in mfiles exactly the same way it works with inline functions, e.g., `plot(x,f(x))` graphs the function given by the mfile `f.m`.

However, the command `ezplot` uses a slightly different call. Remember, for a function given as an inline function, or defined symbolically, the call is `ezplot(f,1,3)`. When the function is given in an mfile `f.m`, the call is `ezplot('f', 1,3)`. Note the single quotes. We shall see this difference often.

`ezplot` is an example of a function mfile that can operate on other functions. These function mfiles have the *name* of a function as an argument in the call. When the function is given as an inline function, the name of the function is `f` or `g`, etc. When the function is given in an mfile, the name of the function is `'f'` or `'g'`, etc. We give two more examples of this type of function mfile in Section 2.3.

2.3   Function functions

MATLAB has a number of routines that operate on functions, called *function functions*. These are function mfiles that generally have function names as well as variables as arguments. We give only a couple of examples that we shall use later.

The root finder `fzero` finds numerical estimates of the roots of an equation $f(x) = 0$. First we define f in an mfile or as an inline function. If f is continuous and changes sign in the interval $[x_0, x_1]$, then there must be a root x_* of $f(x) = 0$ in this interval. When f is defined as an inline function, we can get a numerical estimate of the root with the call `root = fzero(f, [x0, x1])`. If f is defined in an mfile, the call is `root = fzero('f', [x0, x1])`. Note that in the latter case, we use single quotes around `f`.

Example 2.1

The function $f(x) = \sin x - x/2$ changes sign in the interval $[1, 3]$. To find the root of $f(x) = 0$ in this interval we use the following commands:

```
>> f = inline('sin(x) - x/2')
>> root = fzero(f, [1,3])
   root = 1.8955
```

There are many options that can be used with `fzero`. The function function `fzero` is discussed further in Chapter 7. See also the online help.

A second important routine that we shall use is a numerical integrator. If $f(x)$ is given on the interval $[a, b]$, the call `quad8(f,a,b)` makes a numerical estimate of $\int_a^b f(x)dx$. Again, when f is defined in an mfile, we must use single quotes in the call. We shall discuss this numerical integrator more in Chapter 9. Information is available online with `help quad8`.

2.4   Script mfiles

Script mfiles are used to collect a sequence of commands that constitute a program. When we enter the name of the script mfile on the command line, the program will be executed. Here are two examples.

Example 2.2

Suppose that we wish to plot the functions $f_n(x) = x^n \exp(-nx)$ on the interval $[0, 20]$ for $n = 1, \ldots, 10$ on the same graph. We could do this by using the `plot` command and `hold on` over and over again on the command line. However, a better way, which allows us to reproduce the graphs any time, is to write a short program, call it `graphs.m`, that performs this sequence of repeated operations. We shall use the notion of a *for loop*. Here is the script:

```
x = 0:.1:20;
for n = 1:10
   plot(x, x.^n.*exp(-n*x))
   hold on
end
hold off
```

The command `end` is needed to close the loop. To run this script, enter the command `graphs` on the command line. *Do not* enter the command `graphs.m`.

Example 2.3

In this example, we shall use the root finder `fzero` to find the four roots of the equation $f(x) = e^{-x} - \sin(x) = 0$ that lie in the interval $[0, 10]$. Here is a script that allows us to enter estimates for the four roots at run time and then calculates the roots:

```
f = inline('exp(x) - sin(x)')
x = linspace(0, 10, 101);
plot(x, f(x), x, 0*x, 'g')
est = input('enter the 4 estimates
                        as a four vector [*,*,*,*]    ')
for n = 1:4
    root = fzero(f, est(n))
end
```

In plotting the graph, we also plotted the function identically equal to zero. This puts an x axis in green in the figure and makes it easier to see where the roots are located. After plotting the graph, the program waits for the user to enter four numbers in the form of a vector $[a, b, c, d]$.

Entering Comments

In a function mfile or a script mfile that involves several steps, it is very helpful for you, or for another reader, to identify the steps with comment lines. A *comment line* begins with the percent sign, `%`. When a script or function mfile is executed, the comment lines are ignored. For an example of the use of comment lines, see the script mfile `myexp.m` in Section 2.5.

Workspace

An important difference between script mfiles and function mfiles is in the way the workspace is used. In a script mfile, all definitions of variables and calculations are made in a workspace that is accessible from the command line. In Example 2.3, the vector x can be viewed immediately after running the script simply by entering x on the command line.

By contrast, in a function mfile, the variables are not accessible from the command line. A function mfile has its own workspace, independent of the command line workspace. This arrangement allows us to use variable names in a function mfile that are the same as in other function mfiles with no question of confusion. For example, practically every function mfile using functions of one variable calls that variable x.

2.5 MATLAB documents

Saving your work

You may have to stop a MATLAB session before you have finished a project, and you might like to keep the work you have done so far. The mfiles will be kept in your directory for future use. But there may be expressions created on the command line that you wish to keep. This can be done with the command `save`. For example, suppose you have entered some large matrices $\mathbf{A}, \mathbf{B}, \mathbf{C}$ and a symbolic expression `f = a*x^2 + b*x +c` in the course of a computation. In the next session, you do not wish to reenter these matrices or to retype the symbolic expressions. Instead, you can save them to a file, e.g., `hotstuff`, with the command `save hotstuff`. This will save the values of all the variables you have used. If you only want to save the values of $\mathbf{A}, \mathbf{B}, \mathbf{C}$, you can refine the command to `save hotstuff A B C`. At your next session, to retrieve these variables, you would use the command `load hotstuff`.

Saving figures

To save a figure so that you can do further work on it, put the commands that generated the figure in an mfile, e.g., `fig1.m`. When you enter `fig1` on the command line, the figure will be generated.

To print out a figure, click on the "file" button on the upper left of the figure window and select print. You can also type `print` on the command line. This should print out the figure if you are working on a stand-alone machine connected to a printer. If you are working in a network of machines, you may need additional instructions. Ask your system manager for help.

To prepare a figure to be included in another document, give the figure a name, e.g., Fig1, and use the command `print -deps Fig1`. This will save the figure in the form of a Encapsulated PostScript file, `Fig1.eps`, that should be stored in your current working directory. The figure can also be printed out if your machine or system can print out PostScript files. For a list of printing options, enter `help print`.

Preparing MATLAB documents

It is important to be able to present your MATLAB work in a well-organized, readable manner. Here are some instructions to help you do this. We illustrate with an example. Suppose problem 1 in some assignment asks you to sum the power series for e^x with 5 terms, compare with the MATLAB function `exp(x)`, and plot the results on the interval $[-2, 2]$. This would be done with a script mfile, which we shall call `myexp.m`. It would consist of the following sequence of commands:

```
% define the vector of points where the function is to be
% computed and plotted.
x = -2:.2:2;

% the first term in the approx. is set equal to 1.
term = ones(size(x));
y = term;

% add up the terms and store the result in the vector y
for n = 1:4
   term = term.*(x/n);
   y = y + term;
end

% display the results as column vectors
[x', y', exp(x)', (y-exp(x))']
maxerror = max(abs(y - exp(x)))
plot(x,y,x,exp(x), '--')
```

Now when you enter the command `myexp`, you will produce four columns of numbers on the screen, the number "maxerror" on the screen, and a graph in a figure window. To record this program to be turned in, together with the output, you use the `diary` commands. The command `diary` *file name* prepares all the following output, together with any keyboard commands, to be put in a text file that can be edited. The command `diary off` after running the program will actually write into the file. In our case, you would enter the commands

```
>> diary problem1
>> myexp
>> diary off
```

The file `problem1` contains the numerical screen output but not the graph. It looks like this:

```
>> myexp
ans =
    -2.0000    0.3333    0.1353    0.1980
    -1.8000    0.2854    0.1653    0.1201
    -1.6000    0.2704    0.2019    0.0685
       .          .          .          .
       .          .          .          .
       .          .          .          .
```

```
     1.6000     4.8357     4.9530     -0.1173
     1.8000     5.8294     6.0496     -0.2202
     2.0000     7.0000     7.3891     -0.3891


maxerror =
     0.3891
>> diary off
```

Notice that the commands of the program itself are not put into the file `problem1`.
To include the program commands as well, use the command `type` in the sequence

```
>> diary problem1
>> type myexp
>> myexp
>> diary off
```

The command `type` reproduces the code of any MATLAB mfile on the screen.

By editing the file `problem1`, it is now possible to add labels at the tops of
the columns and to add interpretive comments about the results of the calculations.
Comments about the graphs can also be added here, with reference to Figure 1,
Figure 2, etc. Here is the file `problem1`, after editing, with the program inserted at
the beginning and a second page for the graph (see Figure 2.1):

```
                        Problem 1
    This is the program "myexp" used to compute a 5-term
approximation to the exponential function on the interval [-2,2].
    x = -2:.2: 2;
    term = ones(size(x));
    y =  term;
    for n = 1:4
       term = term.*(x/n);
       y = y+term;
    end
    [x', y', exp(x)', (y-exp(x))']
    maxerror = max(abs(y-exp(x)))
    plot(x,y,x,exp(x), '--')
    title('Figure 1. 5-term approx,
                      and true exp(x) (dashed line)')
```

```
    The values of the approximation are put in the vector y and
compared with the MATLAB exponential. Here are the results:
```

```
     x          y        exp(x)     error = y - exp(x)

 -2.0000     0.3333     0.1353        0.1980
 -1.8000     0.2854     0.1653        0.1201
 -1.6000     0.2704     0.2019        0.0685
    .          .          .             .
    .          .          .             .
    .          .          .             .
  1.6000     4.8357     4.9530       -0.1173
  1.8000     5.8294     6.0496       -0.2202
  2.0000     7.0000     7.3891       -0.3891


maxerror =
    0.3891
```

Comments: As we can see from Figure 1 (attached), the 5-term approximation does quite well in the interval [-1, 1]. In fact, from the table, we can see the maximum error over this interval is .0099.

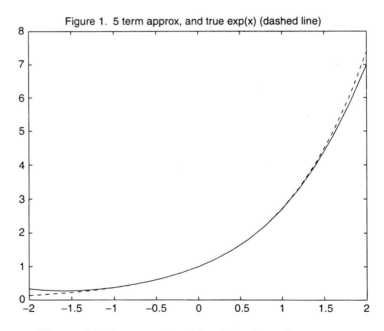

Figure 1. 5 term approx, and true exp(x) (dashed line)

Figure 2.1 Figure produced by the script mfile myexp.

The diary command in Windows

If you are working on a PC with Windows, put the diary into a file with the extension txt. For example,

```
diary myexp.txt
```

A file with the extension txt can be immediately viewed and edited with the Notepad editor.

3

Vectors, Lines, and Planes

In this chapter we study some of the elements of "flat geometry," namely, vectors, lines, and planes. To help with the graphical displays, we have prepared the following mfiles to be used in this chapter.

Prepared mfiles used in this chapter

```
arrow  arrow3  plane
```

3.1 Vectors

Vectors are directed line segments in two- or three-dimensional space, used to represent velocity, acceleration, forces, etc. They have a point of application, a direction, and a magnitude. A vector is usually given in terms of three components, as $\mathbf{v} = [a, b, c]$. The magnitude of the vector is the Euclidean norm of its components, denoted $\|\mathbf{v}\| = \sqrt{a^2 + b^2 + c^2}$. The direction of the vector $\mathbf{v}$ is the unit vector $\mathbf{u} = \mathbf{v}/\|\mathbf{v}\|$. The point of attachment is given by a point $P_0 = (x_0, y_0, z_0)$. Of course, in two-dimensional space, the vectors and points of attachment have only two components.

mfiles arrow, arrow3

The mfiles `arrow.m` and `arrow3.m` are function mfiles that can be used to display vectors in two and three dimensions, respectively. For example, if we want to display vectors $\mathbf{v} = [1, 2]$ and $\mathbf{w} = [-2, 2]$, both attached at the point $P_0 = (2, 6)$, we can use the commands

```
>> P0 = [2 6];
>> v = [1 2];
>> arrow(P0,v)
>> hold on
>> w = [-2 2];
>> arrow(P0,w)
>> arrow(P0,v+w,'r')
>> axis equal
>> hold off
```

Note that in the first two calls to `arrow`, we did not specify the color, and the arrow will be displayed in the default color of blue. In the third call, we added a third argument, `'r'`, which now specifies the color to be red. In this third call we plotted the sum vector $\mathbf{v} + \mathbf{w}$ attached at the same point P_0.

The same kinds of commands work in the same way for the mfile `arrow3.m`.

The next-to-last command, `axis equal`, makes the units on the axes have the same physical length on the screen so that the angles between the vectors are as seen on the screen. The result is shown in Figure 3.1.

The magnitude of a numerical vector $\mathbf{v}$ can be computed in a single MATLAB command, `norm(v)`. If two numerical vectors $\mathbf{v}$ and $\mathbf{w}$ are entered as row vectors (1×3 matrices), their scalar product, $\mathbf{v} \cdot \mathbf{w}$, can be computed with the matrix multiplication `v*w'`. For real vectors, the scalar product is also produced by the command `dot(v,w)`.

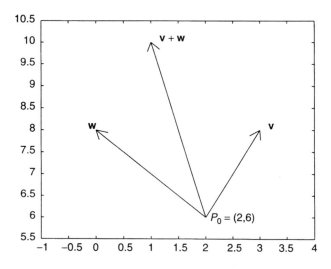

Figure 3.1 Vectors $\mathbf{v} = [1, 2]$, $\mathbf{w} = [-2, 2]$, and $\mathbf{v} + \mathbf{w}$ attached at the point $P_0 = (2, 6)$.

Note that if we use v′ *w, we obtain the 3×3 matrix with entries $v_i w_j$. Finally, the cross product, $\mathbf{v} \times \mathbf{w}$, can be computed with the MATLAB command cross(v,w).

Although the norm command cannot be applied to symbolic vectors, the scalar product and the cross products can be applied to them as shown here:

```
>> syms a b c x y z
>> v = [a b c]
v = [a, b, c]

>> w = [x y z]
w = [x, y, z]

>> cross(v,w)
ans =
[b*z-c*y,c*x-a*z,a*y-b*x]

>> v*w'
ans =
a*conj(x)+b*conj(y)+c*conj(z)
```

In the last result, we see that MATLAB is allowing the possibility that the compo nents of the vectors are complex, and it computes the complex scalar product. If we use the command dot(v,w), the result is a bit strange, conj(a)*x + conj(b)*y + conj(c)*z, with the conjugates on the first factor rather than on the customary second factor.

3.2 Plotting lines in two- and three-dimensional space

If we know two points on the line in two-dimensional space, $P_0 = (x_0, y_0)$ and $P_1 = (x_1, y_1)$, we can plot the line using the plot command:

```
>> plot([x0, x1], [y0,y1])
```

Similarly, if we have points $P_0 = (x_0, y_0, z_0)$ and $P_1 = (x_1, y_1, z_1)$ in three-dimensional space, we use the MATLAB command plot3:

```
>> plot3([x0,x1], [y0,y1], [z0,z1])
```

To plot a polygonal line, joining the five points $P_j = (x_j, y_j, z_j)$, $j = 0, \ldots, 4$, use the commands

```
>> x = [x0, x1, x2, x3, x4];
>> y = [y0, y1, y2, y3, y4];
>> z = [z0, z1, z2, z3, z4];
>> plot3(x,y,z)
```

To plot the points without the lines between them, using an asterisk, for example, replace the last command by `plot3(x,y,z,'*')`. To have the lines between the points and the asterisks at the points, use

```
>> plot3(x,y,z)
>> hold on
>> plot3(x,y,z,'*')
```

Another useful command for graphic display is the `fill` command. The coordinate vectors $x = [x_1, x_2, x_3, x_4]$ and $y = [y_1, y_2, y_3, y_4]$ with $x_4 = x_1$ and $y_4 = y_1$, trace out the vertices of a triangle in the x, y plane. To fill in this triangle in a single color, enter `fill(x,y,'r')`. Here we have taken the color to be red. If the last vertex is not the same as the first vertex, the `fill` command will connect up the last two vertices, thereby filling in a quadrilateral. The same procedure works to fill in any polygon.

The analog in three-dimensional space is to give three vectors of coordinates, as we would to plot a polygonal line. Then the command `fill3(x,y,z,'r')` fills in this polygon in space in red. However, the result can depend on the viewing angle. A consistent result is obtained when the points all lie in the same plane.

Next, we turn our attention to lines described parametrically in two-dimensional and three-dimensional space. If $\mathbf{L} = [a, b, c]$ is a tangent vector to the line and the line passes through the point $P_0 = (x_0, y_0, z_0)$, then its parametric equations are

$$x(t) = x_0 + at, \quad y(t) = y_0 + bt, \quad z(t) = z_0 + ct. \tag{3.1}$$

We can write $P(t) = P_0 + t\mathbf{L}$, and we note that $P(0) = P_0$.

A line given this way is easily plotted. We must first pick a range of values for t. For example, we might want to view the line for $t_1 \leq t \leq t_2$. Then $P(t_1) = (x_0 + at_1, \ y_0 + bt_1, \ z_0 + ct_1)$ and $P(t_2) = (x_0 + at_2, \ y_0 + bt_2, \ z_0 + ct_2)$ are two points on the line. We can use the plotting commands just discussed. After entering values for x_0, y_0, z_0, a, b, c and t_1, t_2, we enter

```
>> x = [x0 + a*t1, x0 + a*t2];
>> y = [y0 + b*t1, y0 + b*t2];
>> z = [z0 + c*t1, z0 + c*t2];
>> plot3(x,y,z)
```

Example 3.1

Another way to plot this same line, which is similar to what we shall do in the next chapter with curves, is to create a vector of t values and then the corresponding coordinate values as follows (here $t_1 = -1$ and $t_2 = 1$):

```
>> t = -1:.01:1;
>> x = x0 + a*t;
>> y = y0 + b*t;
>> z = z0 + c*t;
>> plot3(x,y,z)
>> hold on
% We attach the tangent vector at P0.
>> L = [a b c];
>> P0 = [x0 y0 z0];
>> arrow3(P0,L,'r')
>> hold off
```

The tangent vector attached to the line at P_0 indicates the direction of the line.

3.3 Planes

A two-dimensional plane in three-dimensional space can be determined in several ways. First we suppose we are given a normal direction $N = [a, b, c]$ and a point $P_0 = (x_0, y_0, z_0)$ that the plane is supposed to contain. The plane with normal direction N that contains P_0 is then defined to be the set of points $P = (x, y, z)$ such that the vector $P - P_0$ is orthogonal to N. In coordinates, the plane is given by an equation:

$$\{(x, y, z) : (x - x_0)a + (y - y_0)b + (z - z_0)c = 0\}$$

or

$$ax + by + cz = d \equiv ax_0 + by_0 + cz_0. \tag{3.2}$$

If we are given three points in space, $P_0 = (x_0, y_0, z_0)$, $P_1 = (x_1, y_1, z_1)$, and $P_2 = (x_2, y_2, z_2)$, these three points uniquely define a plane. We can find the equation of this plane in the previous form by finding a normal direction N. Let v be the vector from P_0 to P_1 and w be the vector from P_0 to P_2. Componentwise,

$$v = [x_1 - x_0, \ y_1 - y_0, \ z_1 - z_0] \quad w = [x_2 - x_0, \ y_2 - y_0, \ z_2 - z_0].$$

Now set $N = v \times w$. N will be orthogonal to both v and w. Using this N and P_0 we generate the equation of the plane as before.

If the coefficient $c \neq 0$, i.e., N does not lie in the x, y plane, then we can solve for z in Eq. (3.2) and express the plane as the graph of a function of two variables,

$$z = f(x, y) = z_0 - \frac{1}{c}(a(x - x_0) + b(y - y_0)). \tag{3.3}$$

In Chapter 5 we shall learn how to graph functions of two variables.

In the meantime, we shall turn to another means of describing a plane that is the two-dimensional analog of the parametric equations of a line. The ingredients here are the point P_0 and any two vectors $\mathbf{v}$ and $\mathbf{w}$, which are assumed to be orthogonal to the normal vector $\mathbf{N}$. We also assume that $\mathbf{v}$ is not a scalar multiple of $\mathbf{w}$. These vectors might arise from the knowledge of two other points, P_1 and P_2, in the plane as before. Now for $s, t \in R$, the vector

$$\mathbf{u} = s\mathbf{v} + t\mathbf{w}$$

is said to be a *linear combination* of the vectors $\mathbf{v}$ and $\mathbf{w}$. Any linear combination of $\mathbf{v}$ and $\mathbf{w}$ is also orthogonal to $\mathbf{N}$ because

$$\mathbf{N} \cdot (s\mathbf{v} + t\mathbf{w}) = s\mathbf{N} \cdot \mathbf{v} + t\mathbf{N} \cdot \mathbf{w} = 0.$$

Thus the points P in the plane are given by

$$P = P_0 + s\mathbf{v} + t\mathbf{w} \quad s, t \in R. \tag{3.4}$$

In terms of coordinates, if $\mathbf{v} = [v_1, v_2, v_3]$ and $\mathbf{w} = [w_1, w_2, w_3]$, then $P = (x, y, z)$, where

$$\begin{aligned} x &= x_0 + sv_1 + tw_1 \\ y &= y_0 + sv_2 + tw_2 \\ z &= z_0 + sv_3 + tw_3, \quad s, t \in R. \end{aligned} \tag{3.5}$$

Equations (3.5) are the *parametric equations for the plane*. They are the analog of the parametric equations of the line (3.1).

Equations (3.5) can be used to locate points in a plane. We suppose the values of x_0, y_0, z_0 for P_0 have been entered, as well as the components v_1, v_2, v_3 of $\mathbf{v}$ and the components w_1, w_2, w_3 of $\mathbf{w}$. Then we can display points in the plane with arrows using a short script file (see Figure 3.2):

```
P0 = [x0 y0 z0];
v = [v1 v2 v3];
w = [w1 w2 w3];
for s = -1: .5 :1
    for t = -1: .5 :1
        arrow3(P0, s*v +t*w)
        hold on
    end
end
```

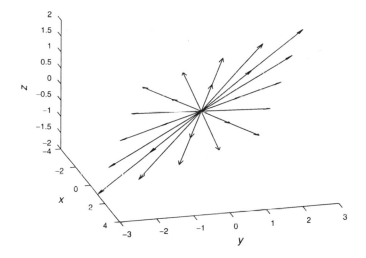

Figure 3.2 Linear combinations of two vectors attached at the origin filling out a plane.

mfile plane

The mfile `plane.m` is a function mfile that graphs a plane using the parametric form. Input information consists of the point P_0 and the normal direction **N**. For example,

```
>> P0 = [1 2 -1];
>> N  = [-5 2 2],
>> plane(P0, N)
```

The plane is shown in Figure 3.3. The normal direction is indicated with a red arrow. Notice that the point P_0 is located in the center of the planar piece, where the normal vector is attached. The width of the planar piece in the horizontal direction is 2 and the width in the other direction is also 2. To get a planar piece of width $2a$ in the horizontal direction and width $2b$ in the other direction, use the longer call

```
>> plane(P0,N,a,b)
```

If two planes have normals $\mathbf{N}_1$ and $\mathbf{N}_2$ that are not multiples of one another, the planes will intersect. The line of intersection will lie in both planes, and hence a tangent vector to this line will be orthogonal to both $\mathbf{N}_1$ and $\mathbf{N}_2$. Thus a tangent direction to the line of intersection will be given by the cross product

$$\mathbf{L} = \mathbf{N}_1 \times \mathbf{N}_2.$$

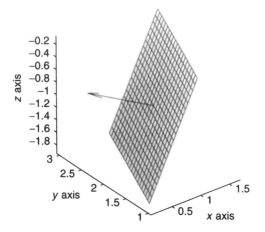

Figure 3.3 Plane through $P_0 = (1, 2, -1)$ with normal $\mathbf{N} = [-5, 2, 2]$.

If P_0 is any point on the line of intersection, then the line is given in parametric form as

$$P = P_0 + t\mathbf{L}.$$

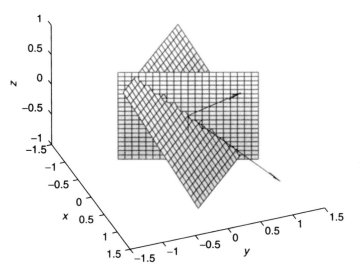

Figure 3.4 Intersection of planes through $P_0 = (0, 0, 0)$ with normals $\mathbf{N}_1 = [-2.5, 1, 1]$ and $\mathbf{N}_2 = [1, 1, 1]$. View is taken with an azimuth of 68 degrees (third quadrant) and an elevation of 32 degrees.

Example 3.2

This script graphs the two planes containing $P_0 = (0, 0, 0)$ with normals $\mathbf{N}_1 = [-2.5, 1, 1]$ and $\mathbf{N}_2 = [1, 1, 1]$. See Figure 3.4. The line of intersection has tangent vector $\mathbf{L} = [0, 3.5, -3.5]$.

```
P0 = [0 0 0];
N1 = [-2.5 1 1];
N2 = [1 1 1];
plane(P0,N1)
hold on
plane(P0,N2)
L = cross(N1, N2);
arrow3(P0,L)
view(68,32)
hold off
```

3.4   Viewing three-dimensional graphs

It is often useful to change the viewpoint in a three-dimensional graph. In MATLAB the viewpoint is specified by a pair of angles, the azimuth and the elevation. Imagine the origin in three-dimensional space translated to the center of the figure. This imposes an x axis, y axis, and z axis in the figure. The azimuth and elevation are determined from these axes. The azimuth is the angle in the x, y plane measured from the negative x axis, with a positive angle being a counterclockwise rotation. The elevation is measured from the x, y plane. The default viewpoint, when the graph first appears, is with an azimuth of -37.5 degrees (in the 4th quadrant) and an elevation of 30 degrees above the x, y plane. To change the viewpoint to an azimuth of 45 degrees and an elevation of 60 degrees, enter view(45,60). At any time to check what the viewpoint is, enter [az,el] = view. In Figure 3.4, the view is from an azimuth of 68 degrees and an elevation of 32 degrees.

Another way to specify the viewpoint is to enter the components of a vector. Imagine this vector attached to the translated origin in the center of the figure. The viewing angle is then down this vector toward the the center of the graph. For example, view([1 2 1]) specifies the viewing angle to be azimuth $153.4 = \arctan(2) + 90$ degrees and elevation $65.9 = \arccos(1/\sqrt{6})$ degrees.

A third way to get a different viewpoint, is to enter the command rotate3d on. Then clicking and holding with the left mouse button, we can rotate the figure. The azimuth and the elevation are displayed on the screen. The command rotate3d off turns the interactive rotate feature off.

Exercises

1. Let $P_0 = (1, 2, 1)$ and $\mathbf{L}_1 = [1, 0, 1]$, $\mathbf{L}_2 = [-1, 2, 0]$, and $\mathbf{L}_3 = [0, 3, 1]$. Plot the three lines through P_0 with these tangent vectors on the same figure using the `hold on` command. Add the tangent vector to each line using the function `arrow3`. Rotate the figure until you are viewing it from an azimuth of 75 degrees and an elevation of 60 degrees. Then label the lines l_1, l_2, and l_3 and the axes.

2. Plot the edges of a regular pentagon inscribed in the unit circle. Then modify the plot to make it a five-pointed star, with the inner points lying on the circle of radius 0.4. See Figure 3.5.

3. Two forces, $\mathbf{F}_1$ and $\mathbf{F}_2$, in the x, y plane are applied to an object placed at the origin. $\mathbf{F}_1$ has a magnitude of 10 pounds and is directed at an angle of 30 degrees above the positive x axis. The force $\mathbf{F}_2$ has a magnitude of 5 pounds and is directed at an angle of 20 below the axis.

a) Find $\mathbf{F}_1$ and $\mathbf{F}_2$. Plot the two forces in the same figure using the `hold on` command and the `arrow` command. Then find the resultant force $\mathbf{F}_1 + \mathbf{F}_2$, and plot it as well in the same figure.

b) Now add a third force, $\mathbf{F}_3 = [0, 1, 2]$. What is the resultant force? Redo the figure in 3D, showing all three forces and the resultant using the `arrow3` command.

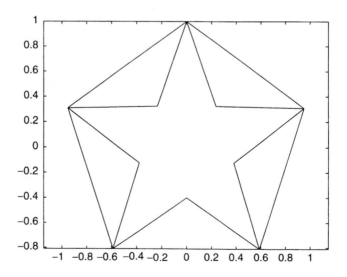

Figure 3.5 Pentagon inscribed in unit circle, with star inscribed in pentagon.

4. Refrigerator Perry, who weighs about 400 pounds, sits on a small three-legged stool. The stool is 2.5 feet high, and the three legs (all of the same length) are fastened at the middle of the bottom of the seat and reach the circle of radius 1 foot on the ground.

a) Use MATLAB to make a three-dimensional diagram of the stool.

b) Compute by hand the magnitude and direction of the force on each leg. Put the force arrows on the diagram with the `arrow3` command.

5. Make an mfile with a "for loop" to graph the family of line segments l_θ through the origin in three-dimensional space

$$l_\theta = \{t(\cos(\theta), \sin(\theta), .7), \quad -1 \le t \le 1\},$$

where $\theta = 2j\pi/n$, $j = 0, \ldots, n$. Choose $n = 24$. What would be the resultant figure as $n \to \infty$?

6. a) Make an mfile with a "for loop" (Section 2.4) to plot the family of lines l_a through the point $P_a = (a, 0, 0)$ with tangent vector $L_a = [0, \cos(a), \sin(a)]$, with $a = j\Delta a$, $\Delta a = \pi/20$ and $j = 0, \ldots, 20$. Add the command `axis equal` to get a better impression. Rotate the view to `view(45,40)`, and label the axes. This is an example of a *ruled surface*, which is a surface constructed from straight lines.

b) Now let a run from $a = 0$ to $a = 4\pi$, with the same Δa. The surface will look like a piece of crepe paper twisted along the x axis. Add the x axis with the command `plot3([0 4*pi], [0 0], 0,0])`. Use the command `axis equal`, and rotate as before. Label the axes.

7. A lever lying in the x, y plane is attached to a vertical shaft that is aligned with the z axis. The end point of the lever is located at $P = (x, y, 0)$. A force $\mathbf{F} = [a, b, c]$ is applied to the lever at P. Let $\mathbf{r}$ be the vector from the origin to P. Then the *torque* at the origin produced by the lever and the force is

$$\mathbf{T} = \mathbf{r} \times \mathbf{F}.$$

For simplicity, take $P = (1, 0, 0)$ so that $\mathbf{r} = [1, 0, 0]$.

a) Use `arrow3` to plot $\mathbf{r}$, $\mathbf{F}$, and $\mathbf{T}$ for various choices of $[a, b, c]$.

b) The z component of the torque is the component that causes rotation about the z axis. Verify that the z component of torque is greatest in magnitude when the force $\mathbf{F}$ is parallel to the xy plane, and perpendicular to the lever.

c) What is the minimum force needed to produce a torque of magnitude 10, that would make the lever rotate in a counterclockwise direction (as seen from above)?

8. a) Let $P_0 = (1, 2 - 1)$, $\mathbf{N}_1 = [2, 1, -1]$, and $\mathbf{N}_2 = [-1, 1, 3]$. Using `plane` and `hold on`, graph both planes through P_0 in the same figure.

b) Let $\mathbf{L} = \mathbf{N}_1 \times \mathbf{N}_2$. Superimpose the line of intersection on the graph of part a). Rotate the figure until the line of intersection is clearly visible. Label the axes.

9. Let $P_1 = (1, 2, 1)$, $P_2 = (2, 5/2, 0)$, and $\mathbf{N} = [1, 1/2, -1]$.

a) Graph the planes through P_0 and P_1 with the same normal $\mathbf{N}$ in the same figure.

b) Let $P_3 = (P_1 + P_2)/2$ and $\mathbf{N}_3 = [0, 1, 1/2]$. Add the graph of this third plane to the previous figure. Rotate the graph so that you can see the intersections clearly. Label the axes.

10. a) Graph the plane that contains the points $P_1 = (1, 0, 0)$, $P_2 = (0, 2, 0)$, and $P_3 = (0, 0, 1)$. Take $P_0 = (P_1 + P_2 + P_3)/3$.

b) To see the portion of the plane that lies in the first octant, let x be the vector of first coordinates of the points P_1, P_2, P_3, let y be the vector of second coordinates, and let z be the vector of third coordinates. Then use the command `fill3(x,y,z,'b')`.

★   **11.** Use the command `plane` and a for loop to construct an array of 8 parallel cooling fins (Figure 3.6). The fins should be square, with side of length 2. They should be centered on the x axis, making an angle of $\pi/6$ with the y, z plane. The perpendicular distance between the fins should be one unit. What is the spacing of the fins in a direction parallel to the x axis? Make the fin at one end of the array contain the origin.

★   **12.** Use the `plane` command to construct a solid bar of length 10 cm and triangular cross section. The cross section should be an equilateral triangle of side 2 cm.

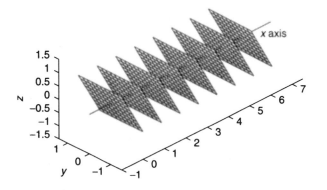

Figure 3.6 Cooling fins for Exercise 11.

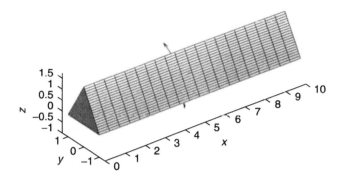

Figure 3.7 Bar of triangular cross section for Exercise 12.

Place the bottom face of the bar in the rectangle $\{0 \le x \le 10, \ -1 \le y \le 1\}$. Choose the points P_0 in the center of each side and the normal direction **N** appropriately. Finally, use the command `fill3` to fill in the ends of the bar. See Figure 3.7.

13. A TV antenna is mounted on a roof. The roof has a slope of 20 degrees from the horizontal. Three guy wires are to be stretched from the antenna to the roof. They are attached at a point 25 feet above the roof, and each is to make an angle of 30 degrees with the pole of the antenna. One of the wires is to be as long as possible and thus must go in the direction in which the roof slopes down. The other

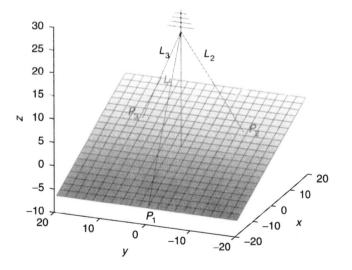

Figure 3.8 Antenna on roof for Exercise 13.

two wires will be shorter and of equal length. When viewed from above, they will each make an angle of 120 degrees with the first wire, one on either side. Introduce coordinates by putting the base of the pole at the origin (0,0,0) and the point where the wires are attached at $(0, 0, 25)$. Let the roof slope up from negative x to positive x. See Figure 3.8.

a) Compute by hand the direction $\mathbf{L}_j$ of the wire L_j, $j = 1, 2, 3$.

b) What is the length of each wire, and what are the coordinates of the points P_1, P_2, P_3 of attachment of each wire on the roof?

c) Graph the roof, the antenna, and the three guy wires.

4

Curves in Space

Prepared mfiles used in this chapter

```
arrow   arrow3   frenet
```

4.1 Parametric representation of curves

Curves in two- and three-dimensional space are often represented as the image of a vector-valued function of a real variable. This is called a *parametric* representation. A parametric representation in two dimensions is provided by two coordinate functions, $x(t)$ and $y(t)$, and the vector-valued function $t \rightarrow (x(t), y(t))$. These curves are very easy to plot with MATLAB.

Example 4.1

A sequence of commands to plot the circle with center at $(1, 3)$ and radius $r = 2$ is

```
>> t = linspace(0, 2*pi, 101);
>> x =   1 +2*cos(t);
>> y =   3 +2*sin(t);
>> plot(x,y)
```

We can also use the feature of the inline function to make such plots. An alternate sequence of commands to graph the circle is

```
>> x = inline('1 + 2*cos(t)');
>> y = inline('3 + 2*sin(t)');
>> t = linspace(0, 2*pi, 101);
>> plot(x(t), y(t));
```

The inline functions x and y defined in the first and second lines are array-smart; when applied to the vector t, they produce the same vectors as in the second and third lines of the previous sequence.

Yet a third way of graphing these two-dimensional curves is possible in MATLAB 5.2 and higher using symbolically defined functions:

```
>> syms t
>> x = cos(t)
>> y = sin(t)
>> ezplot(x,y)
```

In this version of the ezplot command, the default range of the parameter t is $0 \le t \le 2\pi$. If we wish to specify a different range, such as $1 \le t \le 5$, we use the longer command ezplot(x,y, [1,5]).

Similarly, curves in three-dimensional space are represented parametrically by vector-valued functions with three coordinate functions, $t \rightarrow (x(t), y(t), z(t))$. A standard example is the circular helix, which is like a circular coil. One parametric representation is

$$x(t) = \cos t, \quad y(t) = \sin t, \quad z(t) = t/2\pi, \quad 0 \le t \le 4\pi. \qquad (4.1)$$

The graph is shown in Figure 4.1.

To graph curves in three-dimensional space, we use the command plot3. For example, to graph the circular helix in Figure 4.1, we enter

```
>> t = linspace(0, 4*pi, 201);
>> plot3(cos(t), sin(t), t/(2*pi))
```

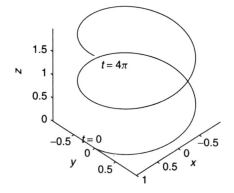

Figure 4.1 Circular helix, rising to height $z = 2$ in two revolutions.

If the coordinate functions were more complicated, we might prefer to define them first as inline functions or as mfiles.

Again, in MATLAB 5.2 and higher, there is a command ezplot3, which is the analog of ezplot. If the coordinate functions x, y, z are symbolically defined functions, we can graph the curve in three-dimensional space with the command ezplot3(x,y,z,[0,4*pi]). We shall describe an additional bell and/or whistle of ezplot3 shortly.

4.2   Tangent vectors and velocity

The notion of the tangent vector to a curve can be motivated by looking at secant lines, much in the way that we find the slope of the tangent to a curve in one dimension. Consider the curve parameterized by $P(t) = (x(t), y(t), z(t))$ near $t = t_0$. Let $P_0 = P(t_0)$. The vectors

$$\frac{P(t) - P(t_0)}{t - t_0} = \left[\frac{x(t) - x(t_0)}{t - t_0}, \frac{y(t) - y(t_0)}{t - t_0}, \frac{z(t) - z(t_0)}{t - t_0} \right]$$

describe the directed line segment from the point P_0 on the curve to the point $P(t)$, multiplied by $(t - t_0)^{-1}$. In Figure 4.2 we show two of these secant vectors, $\mathbf{v}_1 = (P_1 - P_0)/(t_1 - t_0)$ and $\mathbf{v}_2 = (P_2 - P_0)/(t_2 - t_0)$ with $P_1 = P(t_1)$ and $P_2 = P(t_2)$, where $t_0 < t_1 < t_2$.

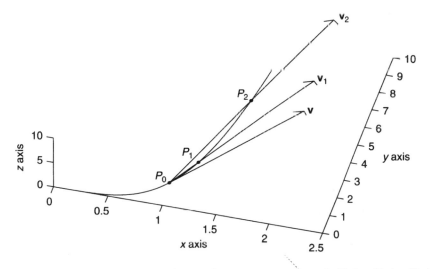

Figure 4.2 Secant vectors $\mathbf{v}_1$ and $\mathbf{v}_2$ and tangent vector $\mathbf{v} = [x'(t_0), y'(t_0), z'(t_0)]$.

If the coordinate functions $x(t)$, $y(t)$, $z(t)$ are differentiable at t_0, the components of the secant vectors converge to $x'(t_0)$, $y'(t_0)$, $z'(t_0)$. The vector

$$\mathbf{v} = [x'(t_0), y'(t_0), z'(t_0)]$$

is a *tangent* vector to the curve at $P(t_0)$. We say *a* tangent vector because any scalar multiple of $\mathbf{v}$ is also a tangent vector. See Figure 4.2. If for some reason we cannot compute the derivatives $x'(t_0)$, $y'(t_0)$, $z'(t_0)$, we can use the components of $(P(t) - P(t_0))/(t - t_0)$ to approximate the derivatives. This is called *numerical differentiation* and is discussed in more detail in Section 4.6.

If $P(t)$ represents the position of a particle at time t, then the tangent vector $\mathbf{v} = [x'(t), y'(t), z'(t)]$ is interpreted as the *velocity* of the particle. The *speed* of the particle is

$$\|\mathbf{v}(t)\| = \sqrt{x'(t)^2 + y'(t)^2 + z'(t)^2}.$$

The *acceleration* of the particle is

$$\mathbf{a} = \frac{d\mathbf{v}}{dt} = [x''(t), y''(t), z''(t)].$$

Example 4.2

We can add tangent vectors at several points along the graph of a curve using the `arrow` and `arrow3` commands. We return to the helix of Figure 4.1, which is parameterized by Eqs. (4.1). The velocity here is $\mathbf{v}(t) = [-\sin t, \cos t, 1/2\pi]$. We can graph the helix, together with tangent vectors at times $t = 0, \pi/4, \pi/2, \ldots, 4\pi$ with the following script:

```
t = linspace(0, 4*pi, 101);
x = inline('cos(t)');
y = inline('sin(t)');
z = inline('t/(2*pi)');
plot3(x(t),y(t),z(t))
hold on
for s = linspace(0, 4*pi, 17);
    p = [x(s),y(s),z(s)];
    v = [-sin(s), cos(s), 1/(2*pi)];
    arrow3(p,v,'r')
end
view(135,40)
```

In lines 2, 3, and 4, we defined the coordinate functions with the inline feature. We did not do the same for the derivatives x', etc., because we used them only once. The result is shown in Figure 4.3.

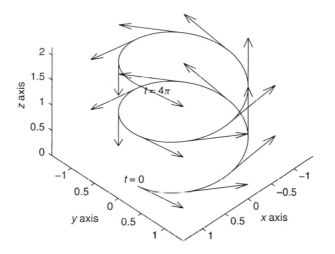

Figure 4.3 Circular helix with velocity vectors.

The command `ezplot3` can be used here, with an additional argument that causes a red bead to move along the curve from initial point to terminal point. The call is

```
>> ezplot3(x,y,z,[0, 4*pi], 'animate')
```

The velocity of the bead varies as the velocity (x', y', z') varies. If we replace the parameter t by $2t$, the bead moves twice as fast.

Often a curve can have sharp corners, or cusps. For example in two dimensions, $P(t) = (x(t), y(t)) = (t^3, t^2)$, for $|t| \leq 1$ (see Figure 4.4), has a cusp at $(0, 0)$. We note here that $\mathbf{v}(0) = [x'(0), y'(0)] = [0, 0]$.

A curve C will not have cusps if it can be parameterized by a vector-valued function $t \to P(t) = (x(t), y(t), z(t))$, $a \leq t \leq b$, such that $x(t)$, $y(t)$, and $z(t)$ are continuously differentiable on $[a, b]$ *and* $\|\mathbf{v}(t)\| > 0$ on $[a, b]$. The latter condition says that if $P(t)$ represents the motion of a particle, the velocity is never zero. If C has a parameterization $P(t)$ that satisfies these two conditions, we say that C is a *smooth* curve. We say that C is *piecewise smooth* if it can be parameterized by a function $t \to P(t)$, $a \leq t \leq b$, such that P is smooth except at a finite number of points t_j and at each t_j, the one-sided derivatives exist. The curve parameterized by $P(t) = (t^3, t^2)$, $|t| \leq 1$, is piecewise smooth. The velocity $\mathbf{v} = [3t^2, 2t]$ vanishes at $t = 0$. Another piecewise smooth curve is the path consisting of the four sides of a square. Here it would be more convenient to use four different functions, one for each side.

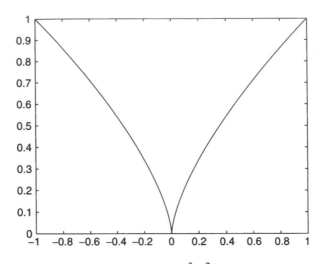

Figure 4.4 Graph of curve $P(t) = (t^3, t^2)$ showing cusp at $t = 0$.

To unify the notation somewhat, we shall adopt the common practice in physics and engineering of identifying the point $P = (x, y, z)$ with the vector attached at the origin with the same components, $\mathbf{r} = [x, y, z]$.

Example 4.3

The cycloid curve is an example of a piecewise smooth curve. It is the path followed by a light attached to the rim of a wheel as it rolls. Let the wheel have radius r. We assume that the wheel is rolling down the positive axis with angular velocity ω. Since the center of the wheel is always located directly over the point of contact with the ground, the center of the wheel moves in a horizontal direction with constant velocity $v = r\omega$. Hence the center of the wheel will be located by the vector $\mathbf{r}_0(t) = [vt, r]$. Since the wheel is rolling in a clockwise fashion, the light on the rim moves clockwise about the center. Its motion about the center is given by $\mathbf{r}_1(t) = [r\cos(\omega t + \delta), -r\sin(\omega t + \delta)]$, where δ is a phase factor determined by the position at time $t = 0$. Hence the position of the light on the wheel rim is given by

$$\mathbf{r}(t) = \mathbf{r}_0(t) + \mathbf{r}_1(t) = [r\omega t + r\cos(\omega t + \delta), r - r\sin(\omega t + \delta)].$$

If we specify that the light at $t = 0$ is to be at the point $(0, 0)$ (the point of contact with the x axis), we must take $\delta = \pi/2$. After applying some trigonometric identities, the equations become

$$\mathbf{r}(t) = [r(\omega t - \sin \omega t), \ r(1 - \cos \omega t)].$$

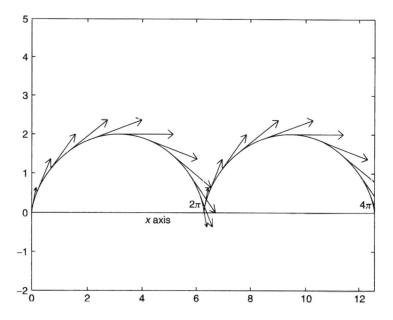

Figure 4.5 Cycloid curve with tangent vectors for wheel of radius 1 with angular velocity $\omega = 1$.

The cycloid curve, together with several velocity vectors, is graphed with the script (see Figure 4.5),

```
% Graph the cycloid curve
rad = 1; omega = 1;
t = linspace(0, 4*pi, 121);
x = rad*(omega*t -sin(omega*t));
y = rad*(1 - cos(omega*t));
plot(x,y)
hold on

% Add the velocity vectors
for s = linspace(0, 4*pi, 25);
   r = [rad*(omega*s -sin(omega*s)),
                          rad*(1-cos(omega*s))];
   u = [rad*omega*(1- cos(omega*s)),
                          rad*omega*sin(omega*s)];
   arrow(r,u,'r')
end
hold off
```

Note that the velocity is zero when $t = 0, 2\pi$, and 4π. These are the points of contact with the ground, and the curve traced out by the light has a cusp at these points.

4.3   Arc length

The length of a line segment from point $P_0 = (x_0, y_0, z_0)$ to point $P_1 = (x_1, y_1, z_1)$ is simply the norm of the vector

$$P_1 - P_0 = [x_1 - x_0, y_1 - y_0, z_1 - z_0]$$

and this is

$$\|P_1 - P_0\| = \sqrt{(x_1 - x_0)^2 + (y_1 - y_0)^2 + (z_1 - z_0)^2}.$$

Similarly the length of a polygonal path Γ with vertices $P_0, P_1, \ldots, P_N$ is the sum of the length of the connecting line segments:

$$\text{length}(\Gamma) = \|P_1 - P_0\| + \|P_2 - P_1\| + \cdots + \|P_N - P_{N-1}\|. \tag{4.2}$$

The fact that we can easily calculate the length of a polygonal path suggests that we could give meaning to the length of a curve C by approximating the curve by a polygonal path Γ and then calculating the length of Γ. Our procedure is to select points $P_0, \ldots, P_N$ on C and then to form the sum (4.2). The polygonal approximation is illustrated in Figure 4.6.

As we choose more and more points on C, the polygonal path Γ becomes a better and better approximation to the curve C. When C is piecewise smooth, the lengths of the approximating polygonal paths converge to a number $l(C)$, which we shall take as the length of C. If C is parameterized by $\mathbf{r}(t)$, $a \leq t \leq b$, with velocity vector $\mathbf{v}(t) = [x'(t), y'(t), z'(t)]$, the length $l(C)$ is given by the arc length integral

$$l(C) = \int_a^b \|\mathbf{v}(t)\| \, dt = \int_a^b \sqrt{x'(t)^2 + y'(t)^2 + z'(t)^2} \, dt.$$

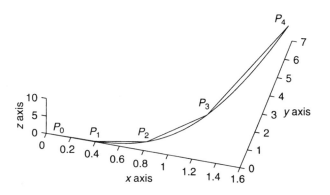

Figure 4.6 Polygonal path Γ approximating curve C.

Because of the square root appearing in $\|\mathbf{v}\|$, there are not many examples where the integral can be computed by elementary means. There are several alternative methods, depending on the goal of our computation.

- If we wish to get a value for the arc length showing the dependence on parameters, for example, the upper limit of integration b, we might try the symbolic integration features of MATLAB.

- If we want only a number for the arc length of a specific curve and we have an expression for the derivatives x', y', z', then we can use a numerical integration routine of MATLAB, such as quad or quadl.

- Finally, if we are given a parameterization $t \rightarrow \mathbf{r}(t)$ but it is difficult to compute the derivatives x', y', z', we can compute the length of an approximating polygonal path.

Example 4.4

Let the curve C be parameterized by $\mathbf{r}(t) = [2t,\ t^2,\ \ln(t)]$, $1 \le t \le 2$. First we approximate the length of C by making a polygonal approximation. Here is the script that calculates the sum (4.2)

```
t = 1:.01:2;
x = 2*t; y = t.^2; z = log(t);
sum = 0;
for j = 1:100
    dx = x(j+1) - x(j);
    dy = y(j+1) - y(j);
    dz = z(j+1) - z(j);
    dr = [dx,dy,dz];
    sum = sum + norm(dr);
end
disp(' this is the length of the polygonal approx
        using 100 segments  ')
sum
```

Now we instead calculate the speed by hand. We have

$$\|\mathbf{v}(t)\| = \sqrt{4 + 4t^2 + (1/t)^2} = 2t + 1/t.$$

The arc length integral is trivial to compute by hand, but we illustrate how to use the numerical integrator quadl. Use the following commands:

```
>> speed = inline('2*t + 1./t')
   Inline function :
   speed(t) = 2*t + 1./t
>> quadl(speed, 1,2)
ans =
      3.6931
```

If speed were defined by an mfile rather than an inline function, the integration command would be quadl('speed', 1,2). Note the single quotes.

Finally we make a symbolic calculation, with the upper limit of integration being a parameter b:

```
>> syms t b
>> r = [2*t, t^2, log(t)]
>> v = diff(r)
v =
[2, 2*t, 1/t]
>> speed = sqrt(v(1)^2 + v(2)^2 + v(3)^2))
speed =
(4 +4*t^2 +1/t^2)^(1/2)
>> int(speed, t, 1, b)
ans =
(((1+2*b^2)^2/b^2)^(1/2)*b^3+
((1+2*b^2)^2/b^2)^(1/2)*b*log(b)-1-2*b^2)/(1+2*b^2)
```

The command pretty(ans) brings this mess into a more intelligible form:

```
>> pretty(ans)

/   2        4     \1/2    /   2        4     \1/2
|4 b  + 4 b  + 1|   3    |4 b  + 4 b  + 1|                    2
|---------------| b  +  |---------------|b log(b)  - 1 - 2 b
|        2      |       |        2      |
\        b      /       \        b      /
--------------------------------------------------------------
                               2
                        1 + 2 b
```

Of course, this last expression can be simplified to $b^2 - 1 + \ln(b)$.

4.4 The geometry of curves

To focus on the geometric aspects of the curve C, independent of the parameterization, we shall factor out the speed, retaining only the changing direction of the velocity. We set

$$\mathbf{T}(t) = \frac{\mathbf{v}(t)}{\|\mathbf{v}(t)\|} = \frac{[x'(t), y'(t), z'(t)]}{\sqrt{x'(t)^2 + y'(t)^2 + z'(t)^2}}. \tag{4.3}$$

$\mathbf{T}$ is the *unit tangent vector*.

The changes in direction of the unit tangent vector as the point $P(t)$ moves along the curve tell us how the curve is twisting and turning in space. Since $\|\mathbf{T}(t)\| = 1$ for all t, we have

$$\mathbf{T}'(t) \cdot \mathbf{T}(t) = 0.$$

Hence

$$\mathbf{N}(t) = \frac{\mathbf{T}'(t)}{\|\mathbf{T}'(t)\|} \tag{4.4}$$

is a unit vector orthogonal to $\mathbf{T}(t)$. $\mathbf{N}(t)$ is called the *principal normal* vector to C at the point $P(t)$. Finally, a third orthogonal vector is defined by

$$\mathbf{B}(t) = \mathbf{T}(t) \times \mathbf{N}(t). \tag{4.5}$$

$\mathbf{B}$ is called the *binormal*. These three mutually orthogonal unit vectors constitute the *Frenet frame* at each point $P(t)$ on the curve where $\mathbf{T}'(t) \neq 0$. Finally we define the scalar function

$$\kappa(t) = \frac{\|\mathbf{T}'(t)\|}{\|\mathbf{v}(t)\|}. \tag{4.6}$$

κ is called the *curvature*.

The unit tangent vector is not too difficult to compute. We can compute $\mathbf{N}$, $\mathbf{B}$, and κ in terms of $\mathbf{T}$, $\mathbf{v}$, and $\mathbf{a}$ with the following expressions:

$$\mathbf{N} = \frac{\mathbf{a} - (\mathbf{a} \cdot \mathbf{T})\mathbf{T}}{\|\mathbf{a} - (\mathbf{a} \cdot \mathbf{T})\mathbf{T}\|} \tag{4.7}$$

The numerator in the expression for $\mathbf{N}$ is the projection of the acceleration $\mathbf{a}$ onto the plane orthogonal to $\mathbf{T}$. Then it can be shown that

$$\mathbf{B} = \frac{\mathbf{v} \times \mathbf{a}}{\|\mathbf{v} \times \mathbf{a}\|} \tag{4.8}$$

and

$$\kappa = \frac{\|\mathbf{v} \times \mathbf{a}\|}{\|\mathbf{v}\|^3}. \tag{4.9}$$

The plane spanned by $\mathbf{T}$ and $\mathbf{N}$, with normal $\mathbf{B}$, is called the *osculating plane*. The circle with radius $1/\kappa$ lying in the osculating plane is the circle that makes the best fit with the curve at that point. *Osculating* comes from the Latin word for "kissing."

From the alternative expression (4.8) for $\mathbf{B}$ we see that both $\mathbf{v}$ and $\mathbf{a}$ lie in the osculating plane. Hence $\mathbf{a}$ can be resolved into a tangential component and a normal component,

$$\mathbf{a} = a_T \mathbf{T} + a_N \mathbf{N}, \tag{4.10}$$

$$a_T = \mathbf{a} \cdot \mathbf{T}, \qquad a_N = \mathbf{a} \cdot \mathbf{N}.$$

a_T has the alternate expression that may be easier to compute,

$$a_T = \frac{d\|\mathbf{v}(t)\|}{dt}.$$

mfile frenet

To help visualize the Frenet frame at points along a curve, we can use the mfile `frenet.m`. It displays the vectors $\mathbf{T}, \mathbf{N}, \mathbf{B}$ at a chosen point along a curve in the figure, and on the screen displays a matrix `frame` of which the first column is $\mathbf{T}$, the second column is $\mathbf{N}$, and the third column is $\mathbf{B}$. The curvature κ is also computed and displayed. To use `frenet`, the coordinate functions $x(t), y(t), z(t)$ must be defined either in mfiles or as inline functions. If they are defined in mfiles, the call is

```
>> frenet('x', 'y', 'z')
```

The user will then be asked to input t values where the Frenet frame is to be computed. This can be done four times. If the coordinates are defined by inline functions, we do not need the single quotes, and the call is

```
>> frenet(x,y,z)
```

Example 4.5

The following sequence of commands was used to produce Figure 4.7.

```
>> x = inline('2*t')
>> y = inline('t.^2')
>> z = inline('t.^3/2')
>> t = 0:.01:2;
>> plot3(x(t), y(t), z(t))
>> axis equal
```

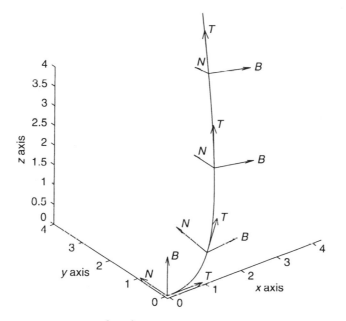

Figure 4.7 Graph of $\mathbf{r}(t) = [2t, \ t^2, \ t^3/2]$, $0 \leq t \leq 2$, showing the Frenet frame at $t = 0, \ 0.8, \ 1.4, \ 1.8$.

```
>> hold on
>> frenet(x,y,z)
 enter value of t
 0
 enter the value of t
.8
 enter the value of t
1.4
 enter the value of t
1.8
```

Notice that at $t = 0$, both **T** and **N** lie in the x, y plane, which in this case is the osculating plane. As we advance along the curve, it is twisting, and the osculating plane rolls in a clockwise direction, as we can see by observing the orientation of **B**.

4.5   Rotations in the plane

Rotations about a point in the x, y plane play an important role in the geometry of planar curves. Let the point (x, y) be represented by the column vector $\mathbf{v} = [x, y]$. This vector can be rotated about the origin by a matrix multiplication. Let θ be an angle, $0 \leq |\theta| < 2\pi$. The following 2×2 matrix is called a *rotation matrix*:

$$\mathbf{R} = \begin{bmatrix} \cos\theta & -\sin\theta \\ \sin\theta & \cos\theta \end{bmatrix}. \tag{4.11}$$

The vector $\mathbf{w} = \mathbf{Rv}$ is the image of $\mathbf{v}$ produced by this rotation. In terms of components,

$$\mathbf{w} = [x\cos\theta - y\sin\theta, \; x\sin\theta + y\cos\theta]. \tag{4.12}$$

The point (x, y) is mapped into the point with the components of $\mathbf{w}$. In particular, note that the point $(1, 0)$ goes into the point $(\cos\theta, \; \sin\theta)$ and the point $(0, 1)$ goes into $(-\sin\theta, \; \cos\theta)$. See Figure 4.8.

What is the image of a curve under rotation? Suppose the curve C is parameterized by $\mathbf{r}(s) = [x(s), y(s)]$, $a \leq s \leq b$. Then the image curve $\tilde{C}$ is parameterized by $\mathbf{q}(s) = \mathbf{Rr}(s)$. To find the components, use Eq. (4.12) on $[x(s), y(s)]$ for each s. We have

$$\mathbf{q}(s) = [x(s)\cos\theta - y(s)\sin\theta, \; x(s)\sin\theta + y(s)\cos\theta]. \tag{4.13}$$

The tangent vectors of C and $\tilde{C}$ are also related. If we differentiate the components of $\mathbf{q}(s)$, we find that a tangent vector to $\tilde{C}$ has components

$$\mathbf{q}'(s) = [x'(s)\cos\theta - y'(s)\sin\theta, \quad x'(s)\sin\theta + y'(s)\cos\theta] \tag{4.14}$$

By referring to Eq. (4.12) we see that $\mathbf{q}'(s) = \mathbf{Rr}'(s)$, which may be shortened to $\tilde{\mathbf{v}}(s) = \mathbf{Rv}(s)$.

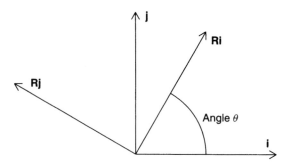

Figure 4.8 Vectors $\mathbf{i} = [1, 0]$ and $\mathbf{j} = [0, 1]$ attached at the origin and rotated through an angle θ.

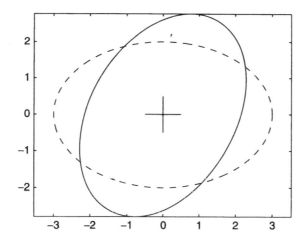

Figure 4.9 Original ellipse (dashed line) and rotated ellipse (solid line), angle of rotation $\theta = \pi/3$.

Example 4.6

We show how to rotate a shape (see Figure 4.9). Let us parameterize the ellipse

$$\frac{x^2}{9} + \frac{y^2}{4} = 1$$

with $\mathbf{r}(s) = [3\cos s, \ 2\sin s]$, $0 \le s \le 2\pi$. The following script rotates this ellipse through an angle θ in a counterclockwise direction when $\theta > 0$. We use Eq. (4.13):

```
theta = input(' enter the rotation angle   ')
s = linspace(0, 2*pi, 101);
x0 = 3*cos(s); y0 = 2*sin(s);
% First we plot the ellipse with no rotation
plot(x0,y0)
hold on
% Now rotate and plot again
x = cos(theta)*x0 -sin(theta)*y0;
y = sin(theta)*x0 + cos(theta)*y0;
plot(x,y);   axis equal
```

4.6 Numerical differentiation

As we saw in the discussion of the tangent vector to a curve, it is possible to approximate the derivative of a function by difference quotients. Note that in the mfile `frenet`, the input functions are only the coordinate functions $x(t)$, $y(t)$, $z(t)$.

However, to compute **T**, **N**, and **B** we need the first and second derivatives of these functions. Since these additional functions are not provided, the derivatives are approximated using numerical differentiation.

Recall from one-dimensional calculus that a function with two continuous derivatives can be approximated by a linear (strictly speaking, affine) function. The linear approximation is given by the first two terms of the Taylor expansion, with an error expressed in terms of the second derivative:

$$f(t+h) = f(t) + f'(t)h + \frac{f''(\xi)}{2}h^2 \tag{4.15}$$

where ξ is some point between t and $t+h$. This equation can be used to approximate the derivative f' in terms of the function values with an error proportional to h,

$$f'(t) = \frac{f(t+h) - f(t)}{h} + E \tag{4.16}$$

where

$$E = -\frac{f''(\xi)}{2}h.$$

The quotient on the right of Eq. (4.16) is called the *forward difference* approximation to $f'(t)$. If we replace h by $-h$ in (4.16) we obtain the *backward difference* approximation to $f'(t)$,

$$f'(t) = \frac{f(t) - f(t-h)}{h} + E.$$

If we assume that f has three continuous derivatives, we can make a longer Taylor expansion,

$$f(t+h) = f(t) + f'(t)h + \frac{f''(t)}{2}h^2 + \frac{f'''(\xi)}{6}h^3. \tag{4.17}$$

Then we can make a more precise expression for the error in the forward and backward difference approximations,

$$f'(t) = \frac{f(t+h) - f(t)}{h} - \frac{f''(t)}{2}h - \frac{f'''(\xi)}{6}h^2. \tag{4.18}$$

and, replacing h by $-h$ in Eq. (4.18),

$$f'(t) = \frac{f(t) - f(t-h)}{h} + \frac{f''(t)}{2}h - \frac{f'''(\eta)}{6}h^2. \tag{4.19}$$

If we add together Eqs. (4.18) and (4.19) and divide by 2, we are taking an average of the forward and backward difference approximations. This average will be more accurate because the $f''(t)h$ term will drop out, leaving error terms involving h^2. The result is called the *centered difference* approximation,

$$f'(t) = \frac{f(t+h) - f(t-h)}{2h} + E. \tag{4.20}$$

The error E of the centered difference approximation is proportional to h^2.

We can extend the Taylor expansion (4.17) to the next term, which involves h^4, if we assume f has four continuous derivatives. Then replacing h by $-h$, and adding the two expansions we see that the terms $f'(t)h$ and $f'''(t)h^3/6$ both drop out. Dividing by h^2, we obtain the *centered difference* approximation for $f''(t)$,

$$f''(t) = \frac{f(t+h) - 2f(t) + f(t-h)}{h^2} + E \tag{4.21}$$

where the error term E is proportional to h^2.

Formulas (4.20) and (4.21) with $h = 10^{-6}$ are used to approximate the components of $\mathbf{v}$ and $\mathbf{a}$ in the mfile `frenet`.

Exercises

1. Let the curve C be parameterized by $P(t) = (1 - \cos t, 1 + 2t + t^2)$.
    a) Calculate by hand a tangent vector to the curve at $P(0)$.
    b) Use MATLAB to compute secant vectors $(P(t) - P(0))/t$ for $t = .2, .1, .05.$ The error in the secant approximation is

$$\frac{P(t) - P(0)}{t} - P'(0).$$

By what factor is the error in each component decreased when t is cut in half?
    c) Plot the curve for $0 \le t \le 1$ and use the `arrow` feature to plot each of these secant vectors as well as the tangent vector computed in part a). Attach all of the vectors to the point $P(0) = (0, 1)$.

★ 2. a) Use the parameterization $x(t) = 2\cos t$, $y(t) = \sin t$, $0 \le t \le 2\pi$, to graph the ellipse $x^2/4 + y^2 = 1$ in the x, y plane.
    b) Calculate by hand the velocity, speed, and acceleration. Where is the speed greatest, and where is the speed the smallest?
    c) Use the command `arrow` to attach the velocity vectors at the points $P(t)$ for the times $t = 0, \pi/3, \pi, 3\pi/2$. See Example 4.2.

d) Calculate the curvature by hand. Then write the speed and curvature as inline functions and graph them together on $[0, 2\pi]$. Where are maximum and minimum values of the speed attained? Same questions for curvature.

e) Use MATLAB to approximate the arc length using a polygonal approximation with 100 segments (see Example 4.4). Then use `quadl` to estimate the arc length integral of the speed. See Example 4.4. In both cases, check your method of calculation on a circle to see if it is working correctly.

3. Let the curve C be parameterized by

$$\mathbf{r}(t) = ((\cos t)^3, (\sin t)^3), \quad 0 \le t \le 2\pi.$$

a) Plot the curve. Does it have cusps? Where do you expect the speed to be zero?

b) Calculate by hand the speed $\|\mathbf{v}\| = \|\mathbf{r}'\|$. Is the speed zero where you expect it to be from your graph of part a)?

c) Calculate by hand the curvature. The formula for a planar curve simplifies to

$$\kappa = \frac{|x''y' - y''x'|}{\|\mathbf{v}\|^3}.$$

Where does κ have singularities? Does this also agree with your graph?

✳   **4.** The curve C of Exercise 3 is piecewise smooth. To make a smooth curve, let us modify $\mathbf{r}$. Let C_δ be parameterized by

$$\mathbf{r}_\delta(t) = \frac{1}{1+\delta}((\cos t)^3 + \delta \cos t, \ (\sin t)^3 + \delta \sin t).$$

a) Plot this curve for values of $\delta = 0.5, 0.2, 0.1$. See how as $\delta \to 0$, C_δ provides a smooth approximation to C.

b) Calculate the speed and curvature of $\mathbf{r}_\delta$ and graph on $[0, 2\pi]$. This is a good opportunity to use the symbolic manipulations of MATLAB and `ezplot`. Let δ and t be symbolic variables and define x and y in terms of them. Then calculate the speed and the curvature in terms of t and δ. Finally, substitute various values of δ and use `ezplot` to graph the speed and curvature together.

5. A projectile with mass $m = 1$ is fired at an angle θ from the horizontal with a speed v_0. The components of its motion are

$$\mathbf{r}(t) = [v_0 t \cos \theta, \ v_0 t \sin \theta - gt^2/2].$$

We are using units of feet and seconds, so $g = 32$ ft/sec/sec. Let $v_0 = 50$ ft/sec.

a) Let t = 0:.01:5. Use the two-dimensional plotting command `plot` to plot the trajectories for θ = 10, 20, 30, 40, 50, 60, 70, 80 degrees. You will need the formula

$$\text{radians} = (\text{degrees}/180) \times \pi.$$

Plot all these curves on the same graph using the command `hold on`. Use the command `axis([0 80 0 50])` to cut them off when they hit the ground.

b) By making further experiments, find that value of θ that yields the maximum range. Then make an analytic calculation to confirm your result. What is the maximum range? Find a formula for the maximum range in terms of the initial speed v_0.

6. Consider a space curve $\mathbf{r}(t) = [3\cos(t),\ 3\sin(t),\ z(t)]$. This is a circular helix of radius $r = 3$ with an unspecified "rise" function $z(t)$.

a) Calculating by hand or symbolically, show that the third component of $\mathbf{a} - (\mathbf{a} \cdot \mathbf{T})\mathbf{T}$ is

$$\frac{9z''(t)}{\|\mathbf{v}(t)\|^2}.$$

b) Using Eq. (4.7) show that when $z(t) = ct$ for some constant c, the vector $\mathbf{N}$ is always parallel to the x, y plane.

c) Now let $z(t) = t^2/2\pi$. Plot the curve for $0 \le t \le 2\pi$. Use the mfile `frenet.m` to add the Frenet frame at the points $P(t)$ for $t = 0, \pi/2, \pi, 3\pi/2$, following Example 4.5 as a model. By looking at the figure, determine if the third component of $\mathbf{N}$ is positive or negative. Does this agree with the calculation of part a)?

d) Same questions for $z(t) = t^{1/2}$, $0 \le t \le 2\pi$, and points $P(t)$ with $t = 0.1,\ \pi/2,\ \pi,\ 3\pi/2$.

7. A plane takes off at time $t = 0$. Its flight path for $0 \le t \le 2\pi$ is the (elliptical) helix

$$x(t) = 2\cos(t), \quad y(t) = 3\sin(t), \quad z(t) = t.$$

a) Plot the curve.

b) Add the Frenet frame at various points. Where is a_N greatest? Where is a_N least?

$\star$ **8.** In this exercise we shall construct a "helix" over the curve C of Exercise 3. Assume $\alpha \ge 0$ and let Γ be the curve parameterized by

$$\mathbf{q}(t) = ((\cos t)^3, (\sin t)^3, \alpha t), \quad 0 \le t \le 2\pi.$$

a) Plot the curve using inline functions and `plot3` for several values of α. Is the curve smooth whenever $\alpha > 0$? Can you explain this by a calculation of the velocity or speed?

b) Now change the z component, and set

$$\mathbf{p}(t) = ((\cos t)^3, (\sin t)^3, \cos(2t)).$$

This curve is known as the *astroid*. Plot this curve using inline functions and `plot3` on $0 \le t \le 2\pi$. Are there cusps present? Make a hand calculation of the velocity. Where is it zero? Calculate the curvature using the symbolic manipulator. Where does it have singularities?

c) After all this work, you should have some fun. Define the components of $\mathbf{p}$ as symbolic expressions. Use the command `ezplot3(x,y,z,'animate')`. Notice how the bead slows down at the cusps.

★   **9.** A 20-foot boat is sitting on the z axis with the stern at $z = 0$ and the bow at $z = 20$. The cross sections of the boat are parabolas in x, $y = (2x/a(z))^2$ for $|x| \le a(z)$, where the coefficient $a(z) = -0.0166z^2 + 0.2245z + 2.25$. There are 21 ribs equally spaced (see Figure 4.10).

a) Plot the rib cross sections for $z = 0, 8, 12, 20$ in the same two-dimensional figure.

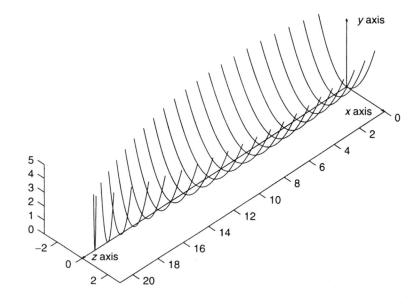

Figure 4.10 Boat hull with 21 ribs.

b) If each of the ribs is to be made of steel tubing, how many feet of tubing are needed for the 21 ribs? Calculating by hand, you can find a formula for the length of each rib. Then add them up using a for loop.

★   **10.** In this exercise, we shall construct the orbit of a moon circling a planet. The planet in turn follows the elliptical orbit parameterized by $\mathbf{r}_0(t) = [3\cos(t),\ 2\sin(t),\ 0]$. The method is similar to that of Example 4.3.

a) Calculate the normal $\mathbf{N}$ of the planet's orbit by hand, and show that the binormal $\mathbf{B} = [0, 0, 1]$.

b) The moon circles the planet in an orbit with radius ρ. It makes 20 complete orbits of the planet in the time the planet completes one elliptical orbit. We use the normal and binormal vectors to describe the moon's position relative to the planet. For convenience, let $\mathbf{n} = -\mathbf{N}$. $\mathbf{n}$ is the normal vector to the ellipse that points to the exterior. Assume that at time $t = 0$, the moon starts at the position $\mathbf{r} = \mathbf{r}_0(0) + \rho\mathbf{n}$. From the point of view of someone on the planet, the motion of the moon is described by the rotation

$$\mathbf{r}_1(t) = \rho\cos(20t)\mathbf{n} + \rho\sin(20t)\mathbf{B}.$$

Write out by hand the components $x(t)$, $y(t)$, $z(t)$ of the combined motion $\mathbf{r}(t) = \mathbf{r}_0(t) + \mathbf{r}_1(t)$.

c) Write a script to graph the orbit of the planet and of the moon with $\rho = 0.2$.

★   **11.** In this exercise we shall describe and graph the motion of a cam and the part that follows the cam. We shall assume the cam in the original position is bounded by a simple closed curve C that goes around the origin. C is parameterized by $s \to [x(s), y(s)]$, where x and y have period 2π. This means that $x(s+2\pi) = x(s)$, and $y(s + 2\pi) = y(s)$ for all s. The cam follower is the part that moves with the cam and lies along the x axis to the right of the cam (see Figure 4.11). We assume that for any angle of rotation θ, the cam follower touches the cam in only one point. Let $l(\theta)$ be the x coordinate of this point of contact. At this point the tangent vector to the rotated cam has a zero x component. From Eq. (4.14), we see that at the point of contact,

$$x'(s)\cos\theta - y'(s)\sin\theta = 0 \tag{4.22}$$

If this equation can be solved for s as a function θ, yielding a function $s(\theta)$, then $l(\theta)$ is given by

$$l(\theta) = x(s(\theta))\cos\theta - y(s(\theta))\sin\theta.$$

a) Let the cam be given by the circle of radius 1, centered at the point $(0.5, 0)$. Parameterize the boundary curve C. Use Eq. (4.13) to write a parameterization of

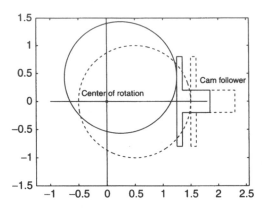

Figure 4.11 Cam and cam follower in original position (dashed line), cam and cam follower with $\theta = \pi/3$ (solid line).

the cam rotated through an angle θ. Note that the circle will not be rotated about its center.

b) Using Example 4.6, write a script that rotates the cam through an angle θ about the origin. Plot the cam in the original position and in the rotated position for $\theta = \pi/3, \ \pi/2, \ \pi$. Label each position and the axes.

c) Solve Eq. (4.22) for s in terms of θ, and find the function $l(\theta)$.

★★   **12.** Continuation of Exercise 11. We consider cams given parametrically by

$$x(s) = c + a \cos s, \quad y(s) = b \sin s.$$

a) Show that Eq. (4.22) becomes $s(\theta) = -\arctan((b/a) \tan \theta)$ and that the expression for $l(\theta)$ becomes

$$l(\theta) = (c + a \cos s(\theta)) \cos \theta - b \sin s(\theta) \sin \theta.$$

b) Write a MATLAB script that takes as input the three parameters a, b, c and graphs the function $l(\theta)$. You will have to adjust values of $s(\theta)$ in the interval $[\pi 2 \le \theta \le 3\pi/2]$ to make $l(\theta)$ continuous.

c) Let $a = 1, \ b = 1/2, \ c = 1/2$. By studying the graph of l, find the angles θ for which the cam follower is at its leftmost point in the cycle. These angles produce the minimum values of $l(\theta)$.

5

Functions of Two Variables

Prepared mfiles used in this chapter

```
qsurf xslice yslice mslice
```

5.1   Defining numerical functions of several variables

Numerical functions of several variables are defined by mfiles and inline functions in the same way as numerical functions of one variable.

Example 5.1

Let $f(x, y) = x + ye^{x^2+y^2}$. This function can be defined in an mfile f.m,

```
function z = f(x,y)
    z = x + y.*exp(x.^2 + y.^2);
```

Notice that we have made the function array-smart by using .* and .^. For graphing purposes it is important that f be able to operate on matrices.

When we define f by an inline function, we must indicate that f is a function of both variables by listing them in single quotes:

```
f = inline('x + y.*exp(x.^2 +y.^2)', 'x', 'y')
```

Mfiles provide a convenient way of writing complicated functions that arise by composition. Suppose the function is $f(x, y) = \sin u \exp(u^2 - v^2)$, where $u = x + y/2$ and $v = \cos(x - y)$. An mfile for this function can be written

```
function z = f(x,y)
    u = x+.5*y;
    v = cos(x-y);
    z = sin(u).*exp(u.^2 - v.^2);
```

5.2   Graphing numerical functions of two variables

In this chapter we discuss graphing a numerical function $f(x, y)$ over a rectangle in the x, y plane, $a \le x \le b$, $c \le y \le d$. For these graphs, the plotting variables are a matrix X of the x coordinates and a matrix Y of the y coordinates. These matrices are constructed as follows. We construct a mesh or grid over the rectangle by selecting a step size in the x direction, Δx, and a step size in the y direction, Δy. Then construct vectors x and y with the commands x = a:delx:b and y = c:dely:d. The next command creates two matrices: [X,Y] = meshgrid(x,y). If $\Delta x = (b - a)/n$ and $\Delta y = (d - c)/m$, then x has length $n + 1$ and y has length $m + 1$. Both matrices X and Y are $(m + 1) \times (n + 1)$. The $m + 1$ rows of X are all equal to x, and the $n + 1$ columns of Y are all equal to y.

Example 5.2

Let $a = -1$, $b = 1$, $c = 0$, and $d = 4$. We choose $\Delta x = \Delta y = .4$ so that $n = 5$ and $m = 10$. The resulting 11×6 coordinate matrices are displayed here.

```
>> x = -1:.4:1; y = 0:.4:4;
>> [X,Y] = meshgrid(x,y)

X =
-1.0000    -0.6000    -0.2000     0.2000     0.6000     1.0000
-1.0000    -0.6000    -0.2000     0.2000     0.6000     1.0000
-1.0000    -0.6000    -0.2000     0.2000     0.6000     1.0000
-1.0000    -0.6000    -0.2000     0.2000     0.6000     1.0000
-1.0000    -0.6000    -0.2000     0.2000     0.6000     1.0000
-1.0000    -0.6000    -0.2000     0.2000     0.6000     1.0000
-1.0000    -0.6000    -0.2000     0.2000     0.6000     1.0000
-1.0000    -0.6000    -0.2000     0.2000     0.6000     1.0000
-1.0000    -0.6000    -0.2000     0.2000     0.6000     1.0000
-1.0000    -0.6000    -0.2000     0.2000     0.6000     1.0000
-1.0000    -0.6000    -0.2000     0.2000     0.6000     1.0000


Y =
     0          0          0          0          0          0
0.4000     0.4000     0.4000     0.4000     0.4000     0.4000
```

0.8000	0.8000	0.8000	0.8000	0.8000	0.8000
1.2000	1.2000	1.2000	1.2000	1.2000	1.2000
1.6000	1.6000	1.6000	1.6000	1.6000	1.6000
2.0000	2.0000	2.0000	2.0000	2.0000	2.0000
2.4000	2.4000	2.4000	2.4000	2.4000	2.4000
2.8000	2.8000	2.8000	2.8000	2.8000	2.8000
3.2000	3.2000	3.2000	3.2000	3.2000	3.2000
3.6000	3.6000	3.6000	3.6000	3.6000	3.6000
4.0000	4.0000	4.0000	4.0000	4.0000	4.0000

Notice that the top row of the Y matrix corresponds to the mesh points on the bottom edge of the rectangle. See Figure 5.1.

Now that we have a mesh, we can graph a function $f(x, y)$ over the rectangle. The graph of f will consist of triples $(x_j, y_i, f(x_j, y_i))$, $i = 1, \ldots, m + 1$, $j = 1, \ldots, n+1$. We shall create an $(m+1) \times (n+1)$ matrix Z, with $Z(i, j) = f(x_j, y_i)$. To be specific, we take

$$f(x, y) = 10x^2 + y^2.$$

We define an inline function f and calculate the matrix Z:

```
>> f = inline('10*x.^2 + y.^2', 'x', 'y')
>> Z = f(X,Y)
```

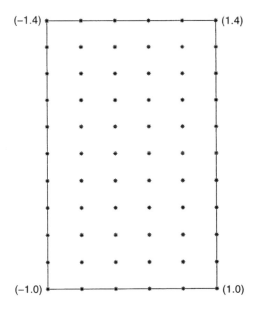

Figure 5.1 Mesh points in the rectangle.

```
Z =
10.0000      3.6000      0.4000      0.4000      3.6000     10.0000
10.1600      3.7600      0.5600      0.5600      3.7600     10.1600
10.6400      4.2400      1.0400      1.0400      4.2400     10.6400
11.4400      5.0400      1.8400      1.8400      5.0400     11.4400
12.5600      6.1600      2.9600      2.9600      6.1600     12.5600
14.0000      7.6000      4.4000      4.4000      7.6000     14.0000
15.7600      9.3600      6.1600      6.1600      9.3600     15.7600
17.8400     11.4400      8.2400      8.2400     11.4400     17.8400
20.2400     13.8400     10.6400     10.6400     13.8400     20.2400
22.9600     16.5600     13.3600     13.3600     16.5600     22.9600
26.0000     19.6000     16.4000     16.4000     19.6000     26.0000
```

The graph is made with the command

```
>> surf(X,Y,Z)
```

We could also combine commands, using surf(X,Y,f(X,Y)). The result is shown on the left in Figure 5.2. The mesh we have used is rather coarse. To make a smoother graph, we use a finer mesh, setting $\Delta x = \Delta y = .2$. The result is on the right in Figure 5.2.

```
>> x = -1:.2:1;
>> y = 0:.2:4;
>> [X,Y] = meshgrid(x,y);
>> surf(X,Y,f(X,Y))
```

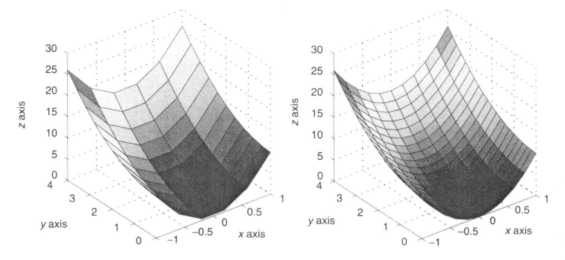

Figure 5.2 Graphs of $f(x, y) = 10x^2 + y^2$, course mesh on the left, finer mesh on the right.

mfile qsurf

This short mfile (quick surf) makes the graphing of numerical functions over a rectangle a bit faster, in that the user does not need to construct a meshgrid. The call is qsurf(f,corners,n) when f is an inline function, and qsurf('f', corners,n) when f is given in an mfile. corners is a four vector, $[a, b, c, d]$ that defines the rectangular domain $R = \{a \leq x \leq b, \; c \leq y \leq d\}$, and n is the number of subdivisions in the mesh in each direction. If the argument n is omitted in the call, the default 50×50 mesh is constructed, which is usually about right. In the case in which n is omitted, the code is essentially this:

```
function out = qsurf(f, corners)
    a = corners(1); b = corners(2);
    c = corners(3); d = corners(4);
    x = linspace(a,b,51);
    y = linspace(c,d,51);
    [X,Y] = meshgrid(x,y);
    Z = feval(f, X,Y);
    surf(X,Y,Z)
```

The only command you have not seen already is feval. It is explained in Chapter 13.

The surf command fills in the facets of the surface and leaves on the straight line segments connecting the points on the graph above the mesh points in the x, y plane. To make a smoother-looking surface, follow the surf command by shading flat or shading interp. The result is shown on the left of Figure 5.3. If we wish to see through the surface, we can use an alternate command, mesh(X,Y,f(X,Y)), which makes a wire frame type of graph, shown on the right of Figure 5.3.

MATLAB uses color to indicate the z coordinate of a surface. You will note that the lowest point on the graph is dark blue, while the highest point is red. This assignment of colors is called a *colormap*. The default colormap is called jet. It is a variant of the colormap hsv (hue-saturation value). There are many different colormaps, suitable for different purposes. To change the colormap to color the graph in shades of gray, use the command colormap(gray). To color the graph in a shading black-red-yellow-white, use colormap(hot). For more information, enter help colormap.

Graphing a numerical function in polar coordinates

Often we may wish to graph a function, not over a rectangle but over a disk. To do this we must make one extra step in the construction of the mesh. The first three

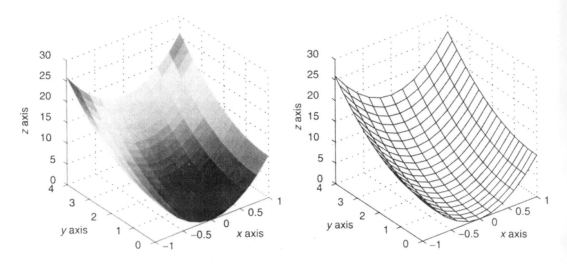

Figure 5.3 Graphs of $f(x, y) = 10x^2 + y^2$, shading flat on the left, wire frame on the right.

lines in the following script create a meshgrid of r, θ coordinates. The fourth and fifth lines convert the r, θ mesh into a curvilinear mesh in the x, y plane.

Example 5.3

The following script shows how to graph the function $f(x, y) = x - 1 + y^2$ over the disk $\{(x - 1)^2 + (y - 3)^2 = 4\}$:

```
% First make a meshgrid in r, theta coordinates
r =  linspace(0,2,21);
theta = linspace(0, 2*pi, 41);
[R,TH] = meshgrid(r,theta);

% Now convert into a curvilinear meshgrid in x,y
% coordinates
X = 1 + R.*cos(TH);
Y = 3 + R.*sin(TH);
Z = X-1 + Y.^2;
surf(X,Y,Z)
hold on

% add the plane z = -5 with the curvilinear
% meshgrid

surf(X,Y,-5+0*Z)
hold off
```

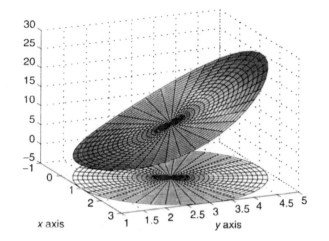

Figure 5.4 Graph of $x - 1 + y^2$ over the disk $(x - 1)^2 + (y - 3)^2 = 4$.

The result is shown in Figure 5.4. The r, θ meshgrid is displayed on the plane $z = -5$.

For a function $g = g(r, \theta)$, we can proceed in the same way.

Example 5.4

We graph $g(r, \theta) = (r/2)\sin(\theta) + r^3 \cos(3\theta)$ over the unit disk, centered at the origin (see Figure 5.5).

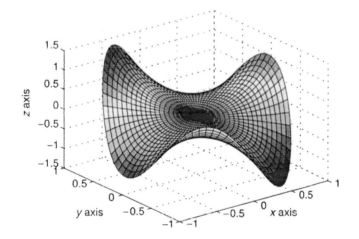

Figure 5.5 Graph of $g(r, \theta) = (r/2)\sin(\theta) + r^3 \cos(3\theta)$ over the unit disk.

```
r = linspace(0,1,21);
theta = linspace(0, 2*pi, 41);
[R,TH] = meshgrid(r,theta);
X = R.*cos(TH);
Y = R.*sin(TH);
Z = .5*R.*sin(TH) + R.^3.*cos(3*TH);
surf(X,Y,Z)
```

5.3   Level curves

The set of points (x, y) where a function takes on a given value is called a *level set* of the function. If the function is $f(x, y)$, we write

$$S_c = \{(x, y) : f(x, y) = c\}.$$

Often these sets are curves in the x, y plane, called *level curves*. In many contexts, the level curves are called *contour lines*. We can visualize how the level curves are generated by slicing the graph of f with a plane $z = c$. The intersection of this plane with the graph is usually a curve in this plane. The level curve is the projection of this curve down on the x, y plane. We illustrate these ideas with the same function we used in Example 5.2: $f(x, y) = 10x^2 + y^2$. In Figure 5.6 (left) horizontal planes at heights $z = 6, 12, 18$ intersect the graph of f. In Figure 5.6 (right) the corresponding level curves are displayed.

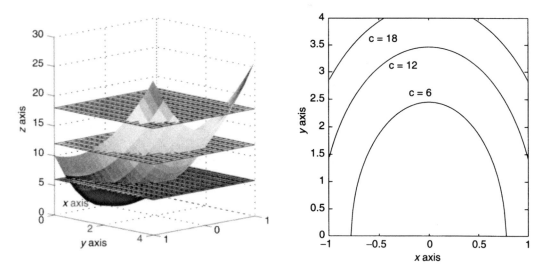

Figure 5.6 On the left, graph of $f(x, y) = 10x^2 + y^2$ intersecting planes of height $z = 6, 12, 18$. On the right, corresponding level curves.

MATLAB has a command that generates level curves of a function. First construct a mesh over some rectangle with $[X,Y]$ = meshgrid(x,y). The basic command contour(X,Y,f(X,Y)) plots several contour lines of f over the rectangle in colors that correspond to the colormap. The basic command can be refined in several ways. contour(X,Y,f(X,Y),n,'k') divides the interval $[\min f, \max f]$ into $n+1$ equal subintervals $\min f < c_1 < c_2 < \cdots < c_n < \max f$ and plots n contour lines corresponding to the n values $c_1 < c_2 < \cdots < c_n$. The last argument, 'k', says that all the contour lines will be plotted in black. Other colors are possible, using the color abbreviations that were listed for the plot command. Finally, we can choose the values of the level curves by substituting a vector of values for the number n. For example, if we wish to plot the level curves for $f(x, y) = 1.5, 2, 2.5, 4$, we define the vector levels = [1.5, 2, 2.5, 4]. These values do not need to be equally spaced. Then make the call contour(X,Y,f(X,Y),levels,'k'). This last method is good for refining a particular area of the contour map where we already know a range of function values. There are further variations on the contour command, which can be found by entering help contour.

We can also make a coloring of the x, y plane that corresponds to the colors in the colormap. This provides another visual key to the behavior of the function. The command is pcolor(X,Y,f(X,Y)). The contour command and this command can be combined very effectively.

Example 5.5

Let

$$f(x, y) = 6e^{-3x^2 - y^2} + x/2 + y. \tag{5.1}$$

We display 10 contour lines, in black, of f in the rectangle $[-2, 2] \times [-2, 3]$ with the following script (see Figure 5.7). The contour lines are superimposed on a pcolor map.

```
x = -2:.05:2; y = -2:.05:3;
[X,Y] = meshgrid(x,y);
f = inline('6*exp(-3*x.^2-y.^2)+.5*x +y','x','y')
Z = f(X,Y);
pcolor(X,Y,Z)
hold on
contour(X,Y,Z,10,'k')
xlabel( ' x axis ')
ylabel( ' y axis ')
hold off
```

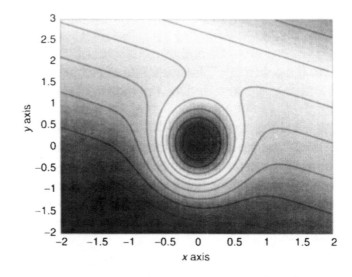

Figure 5.7 Level curves of $f(x, y) = 6 \exp(-3x^2 - y^2) + x/2 + y$.

5.4   Graphing techniques for symbolically defined functions

As noted earlier, MATLAB 5.2 and higher versions include graphing commands for symbolically defined functions of two variables. They include

```
ezsurf
ezmesh
ezcontour
```

A symbolic expression for a function of two variables is given as follows. First we declare x and y as symbolic variables, and then we construct the expression. Here is an example, which includes a graphing command.

```
>> syms x y
>> f = cos(x)*exp(x-y^2)
>> ezsurf(f)
```

When no domain is specified, the graph is made over the default domain $\{-2\pi \leq x, y \leq 2\pi\}$. To specify the domain, use the call ezsurf(f,corners), where, as before, corners is a four vector $[a, b, c, d]$ that defines the rectangle $R = \{a \leq x \leq b, \ c \leq y \leq d\}$. ezmesh and ezcontour work the same way. These commands are especially useful if we take a complicated function and then want to graph some function of the derivatives, e.g., the magnitude of the gradient.

Continuing with the same function and commands, we could write

```
>> fx = diff(f,x); fy = diff(f,y);
>> g = sqrt(fx^2+fy^2)
>> ezsurf(g)
```

5.5 Partial derivatives and the directional derivative

To measure rates of change of a function $f(x, y)$, we consider the restrictions of f to lines parallel to the x axis and parallel to the y axis. This means that we fix one variable, thereby reducing f to be a function of the one remaining variable. For example, the restriction of f to a line parallel to the x axis is $x \to f(x, y_0)$, and the restriction of f to a line parallel to the y axis is $y \to f(x_0, y)$. See Figure 5.8.

Now, we can measure the rate of change of f in a direction parallel to the x axis by differentiating $x \to f(x, y_0)$, and we can measure the rate of change of f in a direction parallel to the y axis by differentiating $y \to f(x_0, y)$. The resulting derivatives are called the *partial derivatives* of f. More specifically we write

$$\frac{\partial f}{\partial x}(x_0, y_0) \quad \text{and} \quad \frac{\partial f}{\partial y}(x_0, y_0)$$

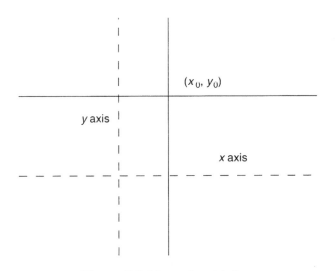

Figure 5.8 Lines of restriction.

for $(d/dx)f(x, y_0)$ evaluated at x_0 and $(d/dy)f(x_0, y)$ evaluated at y_0. The precise definition of the partial derivatives is

$$\frac{\partial f}{\partial x}(x_0, y_0) = \lim_{x \to x_0} \frac{f(x, y_0) - f(x_0, y_0)}{x - x_0} \tag{5.2}$$

and

$$\frac{\partial f}{\partial y}(x_0, y_0) = \lim_{y \to y_0} \frac{f(x_0, y) - f(x_0, y_0)}{y - y_0} \tag{5.3}$$

when these limits exist.

Geometrically, we can see $\partial f/\partial x(x_0, y_0)$ as the slope at $x = x_0$ of the curve that is the intersection of the graph of f and the vertical plane $y = y_0$. On the other hand, $\partial f/\partial y(x_0, y_0)$ is the slope at $y = y_0$ of the curve that is the intersection of the graph of f with the vertical plane $x = x_0$.

mfiles xslice, yslice

The mfiles xslice and yslice slice the graph of a function with a plane $y = y_0$ (xslice) and a plane $x = x_0$ (yslice). We assume that f has already been graphed over a rectangle $[a, b] \times [c, d]$. The call is xslice('f',x,y0) when f is given in an mfile, and xslice(f,x,y0) when f is given as an inline function. x is a vector of x coordinates, ranging from a to b, and y0 is a number. For the other slicer, we use yslice('f',x0,y) or yslice(f,x0,y), and y is a vector of coordinates ranging from c to d.

Example 5.6

Let $f(x, y) = 1 - (x^2 + y^2)/2$. We graph f over the square $\{0 \le x, y \le 2\}$ using qsurf and then slice it with the plane $y = 1$, using the following script. The result is the left figure of Figure 5.9.

```
f = inline('1- .5*(x.^2 +y.^2)','x','y')
qsurf(f, [0 2 0 2]); shading flat
hold on
x = linspace(0,2,51)
xslice(f,x,1)
hold off
```

From Figure 5.9, it is clear that $\partial f/\partial x(1, 1) < 0$ and $\partial f/\partial y(1, 1) < 0$. The right figure was made with yslice.

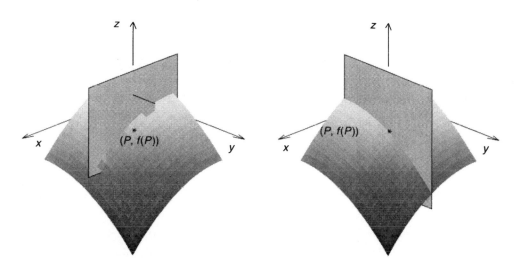

Figure 5.9 On the left, plane $y = 1$ slicing the graph of $1 - .5(x^2 + y^2)$; on the right, plane $x = 1$ slicing the same graph.

The directional derivative

More generally we can compute the rate of change of f at (x_0, y_0) in any direction. Let $\mathbf{u} = [u_1, u_2]$ be a unit vector. The *directional derivative in the direction* $\mathbf{u}$ is defined to be

$$D_{\mathbf{u}} f(x_0, y_0) = \lim_{h \to 0} \frac{f(x_0 + hu_1, \ y_0 + hu_2) - f(x_0, y_0)}{h} \tag{5.4}$$

when this limit exists. We see from Eqs. (5.2) and (5.3) that the partial derivatives $\partial f / \partial x$ and $\partial f / \partial y$ are just the special cases $\mathbf{u} = [1, 0]$ and $\mathbf{u} = [0, 1]$. In fact, once we have computed $\partial f / \partial x (x_0, y_0)$ and $\partial f / \partial y (x_0, y_0)$, we can compute the directional derivative $D_{\mathbf{u}} f(x_0, y_0)$ in any direction $\mathbf{u}$, using the formula

$$D_{\mathbf{u}} f(x_0, y_0) = u_1 \frac{\partial f}{\partial x}(x_0, y_0) + u_2 \frac{\partial f}{\partial y}(x_0, y_0). \tag{5.5}$$

Equation (5.5) shows that $D_{\mathbf{u}} f$ can be expressed as a scalar product of $\mathbf{u}$ with another vector. We define the *gradient vector* of f at the point (x, y) as the vector

$$\nabla f(x, y) = \left[\frac{\partial f}{\partial x}(x, y), \ \frac{\partial f}{\partial y}(x, y) \right]. \tag{5.6}$$

Hence we can write

$$D_{\mathbf{u}} f(x, y) = \nabla f(x, y) \cdot \mathbf{u}. \tag{5.7}$$

It follows immediately that for all directions $\mathbf{u}$,

$$-\|\nabla f(x, y)\| \le D_{\mathbf{u}} f(x, y) \le \|\nabla f(x, y)\|.$$

In particular, the maximum rate of change of f is found in the direction $\mathbf{u} = \nabla f(x, y)/\|\nabla f(x, y)\|$.

mfile mslice

The prepared mfile `mslice` (multislice) displays the behavior of f in the direction $\mathbf{u}$ by intersecting the graph of f with a piece of the vertical plane (parallel to the z axis) $(x - x_0)u_2 = (y - y_0)u_1$. The call is `mslice('f', P)` when f is given in an mfile, and `mslice(f,P)` when f is defined as an online function. $P = (x_0, y_0)$ is the point where the directional derivative is to be calculated.

When the mfile `mslice` is called, the graph of the function is plotted over the square $|x - x_0| \le 1$, $|y - y_0| \le 1$. The script then pauses to allow you to enlarge the window and perhaps rotate the figure to see the surface better. Then enter `return`. A second figure will appear that is a plot in the x, y plane of the point P and three level curves of the function. One of them is the level curve through P. Now move the mouse arrow to a point near P, call it Q, and click. This script will place an arrow in the lower figure from P to Q. In the upper figure, the script shows a vertical plane that stands over the arrow from P to Q. It is part of the larger plane with equation $(x - x_0)u_2 = (y - y_0)u_1$. From the curve of intersection of the graph and the plane you can see whether the function is increasing or decreasing in the direction $\mathbf{u}$, indicated by an arrow in red on the plane. Finally, the script uses finite difference approximations to calculate the directional derivative $D_{\mathbf{u}} f$ at P. The direction $\mathbf{u}$ and $D_{\mathbf{u}} f$ are displayed on the screen. You can see the displays for five directions.

Example 5.7

Let $f(x, y) = x^2 + y$ and fix the point $P = (1, 2)$. We apply the mfile `mslice`. One of the pairs of figures is shown in Figure 5.10. The plane in the upper figure corresponds to the arrow pointing down along the level curve in the lower figure.

```
f = inline('x.^2 +y','x','y')
P = [1 2]
mslice(f,P)
return
u =   [.7809 .6247]
Duf = 2.1864
```

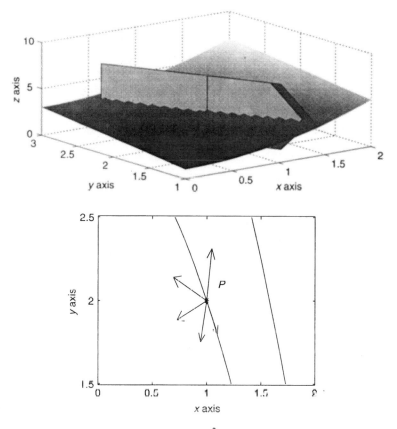

Figure 5.10 Graph of $f(x, y) = x^2 + y$ and slice at $P = (1, 2)$.

```
u  =   [-.2747 .9615 ]
Duf  =  .4121
```

.
.
.

5.6 The gradient vector and level curves

Given a function $f(x, y)$, we define a vector $\mathbf{v} = \nabla f(x, y)$ at each point (x, y) of its domain. This is the gradient vector field. From the definition of the directional derivative it can be seen that $D_{\mathbf{u}} f(x, y) = 0$ when $\mathbf{u}$ is orthogonal to $\nabla f(x, y)$. This leads us to believe, and it can be proved using the chain rule, that at each point (x, y), the gradient vector is orthogonal to the level curve through (x, y).

To display the gradient vector field, we can turn to the MATLAB feature `quiver`. First we write mfiles for each of the partial derivative f_x and f_y, or we define them as inline functions. Then we construct a meshgrid over some rectangular region. Finally we call `quiver(X,Y,fx(X,Y),fy(X,Y))`. The command `quiver` will place an arrow at each point of the mesh. If there are too many points in the x, y mesh, the arrows will overlap and be hard to distinguish. To display both level curves and gradient vectors, we may need to use a coarse mesh for the arrows and a finer mesh for the level curves.

Example 5.8

Let $f(x, y) = xy - x^3/3$. Then $f_x(x, y) = y - x^2$ and $f_y = x$. We shall display the gradient vector field and the level curves of f over the square $[-2, 2] \times [-2, 2]$. See Figure 5.11.

```
f = inline('x.*y - (x.^3)/3', 'x', 'y')
fx = inline('y - x.^2', 'x', 'y')
fy = inline('x', 'x', 'y')
x = -2:.05:2; y = x;
% this is the fine mesh for the level curves.
[X,Y] = meshgrid(x,y);
Z = f(X,Y);
% We choose the level curves
levels = [-6:.5:6];
contour(X,Y,Z,levels)
hold on
xx = -2:.2:2; yy = xx;
% This is the coarse mesh for the arrows
[XX,YY] = meshgrid(xx,yy);
U = fx(XX,YY); V = fy(XX,YY);
quiver(XX,YY,U,V)
axis equal
```

In the next example we use the symbolic tools of MATLAB, in particular the graphing tools `ezsurf` and `ezcontour` of MATLAB 5.2 and higher. This example comes from economics.

Example 5.9

In a monopolist situation, the revenue a firm realizes from the sale of a product or service depends on the production level z. The revenue is an increasing function of z but tends toward an asymptote as the market becomes saturated. Let us suppose that the revenue function $R(z) = z^2/(1 + z^2)$. The production level in turn depends on

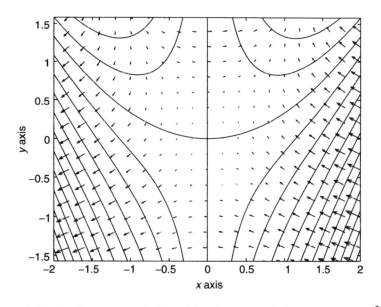

Figure 5.11 Gradient vector field and level curves of $f(x, y) = xy - x^3/3$.

the amount of capital invested, x, and the amount of labor employed, y. A typical production model is the Cobb–Douglas production function $z = Kx^\alpha y^\beta$, where $\alpha + \beta = 1$. Let us suppose $\alpha = \beta = .5$ and $K = 1$. Finally, there is a cost function $C(x, y)$, and we will assume $C(x, y) = .04x + .06y$. Then the profit function

$$\pi(x, y) = R(z(x, y)) - C(x, y).$$

We shall use the symbolic tools of MATLAB to graph the profit function $\pi(x, y)$ and $\pi_x(x, y)$, the marginal profit with respect to capital:

```
>> syms x y
>> z = (x^.5)*(y^.5);
>> revenue = (z^2)/(1+z^2);
>> cost = .04*x + .06*y;
>> profit = revenue - cost;
>> profitx = diff(profit, x);
>> ezsurf(profit, [0 6 0 4])
>> ezcontour(profit, [0 6 0 4])
>> ezsurf(profitx, [0 6 0 4])
>> view, [10, 40]
```

In Figure 5.12 we display the graphs produced by this sequence of commands. From an inspection of the contours of the profit function, we suspect that there is probably a maximum of the profit function near $(x, y) = (3, 2)$.

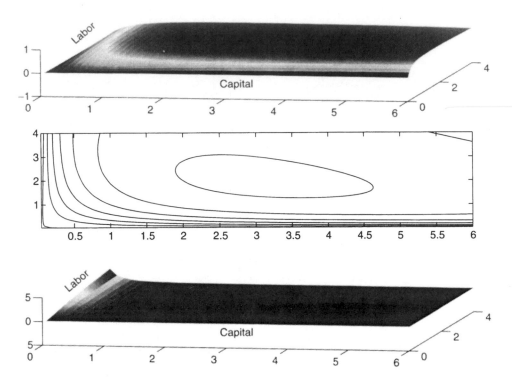

Figure 5.12 On the top, graph of profit function $\pi(x, y)$ of Example 5.9; middle, contours of $\pi(x, y)$; bottom, graph of $\pi_x(x, y)$.

5.7 The tangent plane approximation

It is often important to be able to approximate complicated, nonlinear functions by simpler linear functions, at least locally. This can be done by the tangent plane approximation.

Let $f(x, y)$ be a function and (x_0, y_0) be a point in the domain of f. Let $P_0 = (x_0, y_0, f(x_0, y_0))$ be the point of the graph of f over (x_0, y_0). We shall approximate f by a linear (strictly speaking, affine) function in a region close to (x_0, y_0).

The curve of intersection of the vertical plane $y = y_0$ with the graph of $f(x, y)$ can be parameterized by

$$t \rightarrow \mathbf{r}(t) = (x_0 + t, y_0, f(x_0 + t, y_0)).$$

$\mathbf{r}(0) = P_0$ and the tangent vector to the curve at this point is

$$\mathbf{v} = \mathbf{r}'(0) = [1, 0, f_x(x_0, y_0)].$$

Similarly, the curve of intersection of the plane $x = x_0$ is

$$t \rightarrow \mathbf{q}(t) = (x_0, y_0 + t, f(x_0, y_0 + t)),$$

with tangent vector

$$\mathbf{w} = \mathbf{q}'(0) = [0, 1, f_y(x_0, y_0)].$$

The plane containing P_0 and the vectors $\mathbf{v}$ and $\mathbf{w}$ is given in parametric form by

$$(x, y, z) = P_0 + s\mathbf{v} + t\mathbf{w}, \qquad s, t \in R.$$

Using the chain rule it can be shown that if $\mathbf{p}(t)$ is any curve on the graph with $\mathbf{p}(0) = P_0$, the tangent vector $\mathbf{p}'(0)$ lies in this plane. For this reason the plane spanned by $\mathbf{v}$ and $\mathbf{w}$ is called the *tangent plane* to the graph of f at the point P_0 (see Figure 5.13).

A normal vector to the tangent plane is

$$\mathbf{N} = \mathbf{v} \times \mathbf{w} = [-f_x(x_0, y_0), -f_y(x_0, y_0), 1].$$

The equation for the tangent plane, which is

$$(P - P_0) \cdot \mathbf{N} = 0,$$

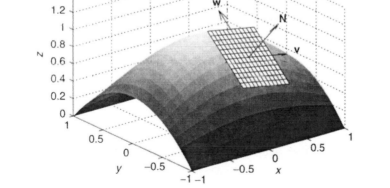

Figure 5.13 Tangent vectors and tangent plane to the graph of $f(x, y) = (1 - y^2)\cos x$ over the point $(x_0, y_0) = (.2, -.4)$.

can be rewritten as

$$z = f(x_0, y_0) + f_x(x_0, y_0)(x - x_0) + f_y(x_0, y_0)(y - y_0). \tag{5.8}$$

For points (x, y) near (x_0, y_0), the tangent plane provides an approximation to the graph of f. If f has continuous second order partial derivatives, we can show that for (x, y) close to (x_0, y_0),

$$f(x, y) = f(x_0, y_0) + f_x(x_0, y_0)(x - x_0) + f_y(x_0, y_0)(y - y_0) + E, \tag{5.9}$$

where $|E| \leq K \max\{|x - x_0|^2, |y - y_0|^2\}$. K is a constant determined by the second derivatives of f. The right-hand side of this equation is the first three terms of the Taylor expansion of f at the point (x_0, y_0).

Example 5.10

We find the tangent plane approximation to $f(x, y) = (1 - y^2) \cos x$ at the point $(x_0, y_0) = (.2, -.4)$. The partial derivatives are $f_x(x, y) = -(1 - y^2) \sin x$ and $f_y(x, y) = -2y \cos x$. Hence the tangent plane to f at $P_0 = (.2, -.4, f(.2, -.4))$ is

$$\begin{aligned} z = l(x, y) &= f(.2, -.4) + f_x(.2, -.4)(x - .2) + f_y(.2, -.4)(y + .4) \\ &= .8233 - .1699(x - .2) + .7841(y + .4), \end{aligned}$$

which has normal vector

$$\mathbf{N} = [-f_x(x_0, y_0), -f_y(x_0, y_0), 1] = [.1699, -.7841, 1].$$

Now we graph f over the square $\{-1 \leq x, y \leq 1\}$ and attach the tangent plane. We graph the tangent plane over the smaller square $\{|x - .2|, |y + .4| \leq .5\}$, and use a coarser mesh to make it more visible.

```
>>   f = inline('(1-y.^2).*cos(x)', 'x', 'y')
>>   l = inline('.8233 - .1699*(x-.2) + .7841*(y+.4)', 'x', 'y')
>>   qsurf(f, [-1,1 -1 ,1])
>>   hold on
>>   qsurf(l, [-.3, .7, -.9, .1], 10)
>>   hold off
```

Figure 5.13 shows the graph of f with the approximating tangent plane and its normal vector $\mathbf{N}$.

Let Q_h be the square centered at (x_0, y_0), $Q_h = \{|x - x_0|, |y - y_0| \leq h\}$. We want to estimate the error E in Eq. (5.9) over Q_h and see how it depends on h. We estimate the maximum of $|E|$ over Q_h, with $h = .2$, with the commands

```
>> x = linspace(0, .4, 101);
>> y = linspace(-.6, -.2, 101);
>> [X,Y] = meshgrid(x,y);
>> E = f(X,Y) - l(X,Y);
>> max(max(abs(E)))
ans = .0619
```

The double `max` command first finds the maximum in each column of a matrix and then finds the maximum over these column maxima. If we repeat this estimate over Q_h with $h = .1$, we get the answer .0156. Thus by cutting h in half, we reduce the maximum error by a factor of approximately one-fourth, as predicted by the statement following Eq. (5.9).

5.8   More about colormaps

A colormap is a way of assigning colors to the graph of a function. Typically a colormap is a spectrum of colors. The default colormap of MATLAB 5.0 is `jet`, which ranges from dark blue to bright red, going through shades of green, yellow, and orange. When we use the command `surf`, the colors are assigned to the graph with the lowest part of the graph colored deep blue, and the highest part colored bright red. To see the range of colors in the colormap and the assignment of numerical values in the z coordinate, enter the command `colorbar`. The colormap color assignment allows us to use color to pick out the maximum and the minimum points of the function over the mesh

Another colormap is `gray`. We switch colormaps with the command `colormap(gray)`. In this colormap, the lowest point on the graph is black, and the highest point is white, with shades of gray in between. `colorbar` will display the shades and the numerical range.

So far, we have used the colormap to indicate the height of the graph above the lowest point on the graph. However, we can also use the colormap to display other features of the graph. This can be done with a fourth argument to the `surf` command. Suppose that `z = f(X,Y)` for some function f and some meshgrid `[X,Y]` and that we wish to indicate on the graph where ∇f is large and where it is small. With f_x and f_y given in mfiles or as inline functions, we set

$$W = sqrt(fx(X,Y).^2 + fy(X,Y).^2)$$

Then the command `surf(X,Y,Z,W)` will assign the colormap colors according to the values of $\|\nabla f(x, y)\|$.

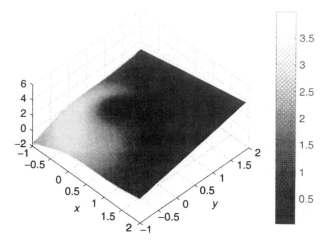

Figure 5.14 Graph of $f(x, y) = x + y + 3 \exp(-x^2 - y^2)$, with colormap gray determined by the magnitude of ∇f.

Example 5.11

Let $f(x, y) = x + y + 3 \exp(-x^2 - y^2)$ so that $f_x = 1 - 6x \exp(-x^2 - y^2)$ and $f_y = 1 - 6y \exp(-x^2 - y^2)$. In this example, the colormap is determined by the magnitude of the gradient. See Figure 5.14.

```
[X,Y] = meshgrid(-1:.1:2);
f  = inline('x + y + 3*exp(-x.^2 -y.^2)', 'x', 'y')
fx = inline('1 - 6*x.*exp(-x.^2 -y.^2)', 'x', 'y')
fy = inline('1 - 6*y.*exp(-x.^2 -y.^2)', 'x', 'y')
Z = f(X,Y);
W = sqrt(fx(X,Y).^2 + fy(X,Y).^2);
surf(X,Y,Z,W); colormap(gray); shading interp;
```

With this colormap, the critical points, where $\nabla f = 0$, are in the darkest regions.

5.9   Cutting off a graph

Often when graphing functions in two and three dimensions, we run into the situation where the function takes on very large values, either positive or negative, in a small region. When this happens the vertical scale of the graph is such that most of the detail is lost. Here are some ways to "cut off" the function values to get a graph that shows more detail.

In two-dimensional graphs, we can use the axis command.

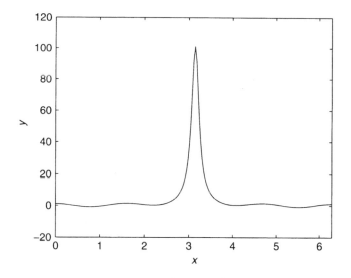

Figure 5.15 Graph of $f(x) = \cos x + [(x - \pi)^2 + .01]^{-1}$ with vertical scale of $-20 \leq y \leq 120$.

Example 5.12

Let $f(x) = \cos x + [(x - \pi)^2 + .01]^{-1}]$. This function peaks at $x = \pi$. If we graph it on the interval $0 \leq x \leq 2\pi$, we lose all the oscillatory detail. See Figure 5.15.

Now we cut off the graph with the command `axis([0 2*pi -2 4])`. The result is shown in Figure 5.16.

Cutting off graphs in three dimensions can also be accomplished with the analogous version of the axis command, but the results are not as successful. In particular, the colormap still uses the maximum and minimum values of the function, so the cut-off graph may be all the same color.

Another way is possible that gives better results. This approach uses the characteristic function of a half interval. Remember that the function described by the expression `(z < 2)` is 1 when the inequality is true and 0 when it is false.

Example 5.13

Let

$$
\begin{aligned}
f(x, y) &= \cos(\sqrt{x^2 + y^2}) + 1/(x^2 + y^2 + .01) \\
&= \cos(r) + 1/(r^2 + .01).
\end{aligned}
$$

We graph f over the square $[-2\pi, 2\pi] \times [-2\pi, 2\pi]$ with the commands

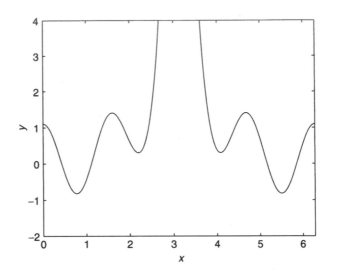

Figure 5.16 Graph of $f(x) = \cos x + [(x - \pi)^2 + .01]^{-1}$ cut off to vertical interval $-2 \le y \le 4$.

```
>> f = inline('cos(sqrt(x.^2 + y.^2)) +
                     (x.^2 +y.^2 + .01).^(-1)', 'x', 'y')
>> [X,Y] = meshgrid(-2*pi:.1:2*pi);
>> Z = f(X,Y);
>> surf(X,Y,Z);
```

The result is shown in Figure 5.17.

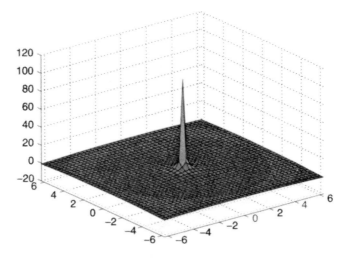

Figure 5.17 Graph with sharp peak at $(0, 0)$ and no detail.

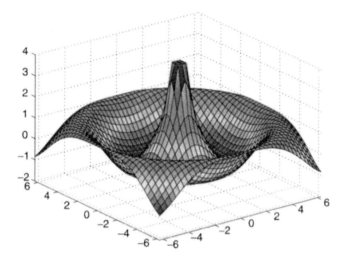

Figure 5.18 Same graph as in Figure 5.17, but cut off to show detail.

Now we cut this graph off at the height $z = 4$ using the characteristic function multiplied times the matrix Z.

```
>> W = (Z-4 < 0).*(Z-4) +4;
>> surf(X,Y,W)
```

The matrix W is obtained from Z first by shifting the level $z = 4$ to the level $z = 0$ by the operation $Z \rightarrow Z - 4$. Then multiplication by the characteristic function sets each positive element of $Z - 4$ to zero. Finally, we shift back to the original level by adding on the constant 4 to each element. The result of this sequence of operations is to set all of the elements of Z that are greater than or equal to 4 as equal to 4. This gives the graph the flat top. See Figure 5.18.

5.10   The subplot command

Often we want to put several graphs in the same figure. This can also save on printing costs. MATLAB accomplishes this with the command subplot. We think of the figure rectangle as broken up into a collection of smaller rectangles by drawing vertical and horizontal lines. Suppose we want to combine six graphs in the same figure, arranging them in three rows of two each. We think of this as a 3×2 array, and we number the smaller rectangles 1,2 across the first row, 3,4 in the second row, and 5,6 in the third row. The command subplot has three arguments. The first two are the dimensions of the array, in this case 3×2. The third argument is the number of the subrectangle. Thus to plot in the first subrectangle in the second row, we use the command subplot(3,2,3).

Example 5.14

Let

$$f(x, y) = ce^{-(x-1)^2 - 2y^2} + (1 - c)e^{-(x+1)^2 - y^2}.$$

We display the graphs of f for six different values of c. The result is shown in Figure 5.19.

```
f = inline('c*exp(-(x-1).^2 -2*y.^2) +
    (1-c)*exp(-(1+x).^2 -y.^2)', 'x', 'y', 'c')

[X,Y] = meshgrid(-2:.2:2);
```

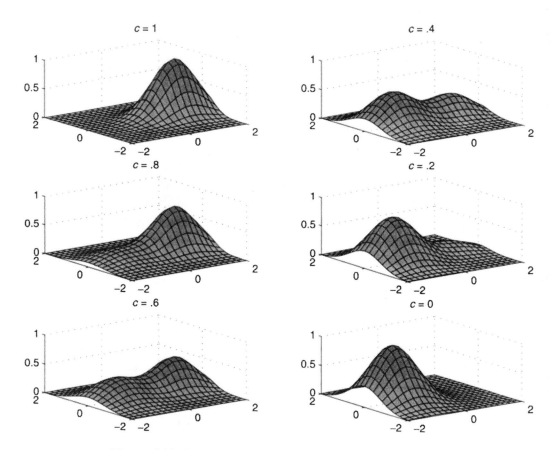

Figure 5.19 Example of the use of the subplot command.

```
subplot(3,2,1)
    surf(X,Y,f(X,Y,1))
    title('c = 1')
subplot(3,2,2)
    surf(X,Y,f(X,Y,.4))
    title('c = .4')
subplot(3,2,3)
    surf(X,Y,f(X,Y,.8))
    title('c = .8')
subplot(3,2,4)
    surf(X,Y,f(X,Y,.2))
    title('c = .2')
subplot(3,2,5)
    surf(X,Y,f(X,Y,.6))
    title('c = .6')
subplot(3,2,6)
    surf(X,Y,f(X,Y,0))
    title('c = 0')
colormap(cool)
```

Exercises

1. Following Example 5.2, set up a mesh on the square $\{-2 \le x, y \le 2\}$, with $\Delta x = \Delta y = .1$. Graph each of the following functions over this square. For each function, make a short script mfile that defines the function as an inline function, constructs the mesh, and graphs the function. Label the axes. Then try the command qsurf on one of them with different values for the mesh parameter n.
   a) $f(x, y) = x^2 - y^2$.
   b) $f(x, y) = \sin(x + y)$.
   c) $f(x, y) = \cos(x^2 + y^2)$.

2. Graph the function $f(r, \theta) = r \sin(2\theta)$ over the disk $\{x^2 + y^2 \le 4\}$. Follow Example 5.4.

3. Write an mfile g.m for the function

$$g(x, y) = 2e^{-(x-1)^2-(y-1)^2} + 1.8e^{-5(x+1)^2-3y^2} - e^{-2(x-1)^2-3(y+.5)^2}.$$

a) Use the meshgrid for the square $\{-2 \le x, y \le 2\}$ that you made in Exercise 1. Make a graph of g. From the figure you can see that the graph of g has two hills and one valley. To see what the level curves will look like, enter hold on and then surf(X,Y,0*X +1). You will see the plane of height 1 intersecting the graph of g. The curve of intersection, projected down in the x, y plane, is the level curve $g(x, y) = 1$. Enter hold off. Label the axes.

(b) To see the contours in the x, y plane, use the contour command, as in Example 5.5, with 20 contour lines. Use the contour map to estimate the coordinates of the two peaks and of the bottom of the valley.

4. Write an mfile, or define as an inline function, the function

$$h(x, y) = (x^2 - y^2)/(x^2 + y^2).$$

a) Graph h on the same square as in Exercise 1. From the graph determine the limits

$$\lim_{x \to 0} h(x, 0) \qquad \text{and} \qquad \lim_{,y \to 0} h(0, y).$$

b) Now use the contour command. What are the level curves of h? Show that for each $a \in R$, the line $y = ax$ is a level curve of h. What is the limiting value of h along this line as $(x, y) \to (0, 0)$? Is h continuous at $(0, 0)$?

5. Let

$$f(x, y) = \frac{x}{2} + e^{-4(x-1)^2 - (y-1)^2}.$$

Write an mfile f.m for f. We shall compute numerical approximations to the partial derivatives of f using the forward difference approximation (see Section 4.6).

$$f_x(a, b) \approx \frac{f(a + \Delta x, b) - f(a, b)}{\Delta x}$$

$$f_y(a, b) \approx \frac{f(a, b + \Delta y) - f(a, b)}{\Delta y},$$

with $\Delta x = \Delta y = .01$. Thus an approximation to the partial derivatives at $(a, b) = (1, 1)$ can be computed by

```
>> fx = (f(1.01,1) - f(1,1))/.01
>> fy = (f(1,1.01) - f(1,1))/.01
```

Compare these approximate values for f_x and f_y with the exact values computed by hand. How does the error in this approximation decrease as Δx and Δy get smaller? If we replace Δx by $\Delta x/10$, does the error decrease by a factor of 10?

6. We continue with the function $f(x, y)$ of Exercise 5. Graph f on the square $\{0 \le x, y \le 2\}$ with the commands

```
qsurf(f, [0 2 0 2])
colormap(gray); shading interp
```

a) Use the mfile `xslice` to slice the graph of f with a vertical plane parallel to the x axis with equation $y = 1$. If necessary, rotate the figure so that you can clearly see the curve of intersection. For what x is $f_x(x, 1) > 0$? For what x is $f_x(x, 1) < 0$? From the concavity of this curve determine where $f_{xx}(x, 1) > 0$ and where $f_{xx}(x, 1) < 0$.

b) Now use the mfile `yslice` to slice the graph of f with the plane $x = 1$. From the curve of intersection, determine where $f(1, y)$ is increasing and where decreasing. From the concavity of this curve, determine where $f_{yy}(1, y) > 0$ and where $f_{yy}(1, y) < 0$.

7. Let $f(x, y) = x|y|$.
a) Graph this function on the square $[-1, 1] \times [-1, 1]$.
b) Use the mfile `xslice` at $y = -.5, 0, .5$. Verify that $f_x(x, y)$ exists. Compute $f_x(x, y)$ by hand.
c) Use the mfile `yslice` at $x = -.5, 0, .5$. By looking at the slopes of the line of intersection of the graph with the slice, deduce that $f_y(x, y) = x$ for $y > 0$ and $f_y(x, y) = -x$ for $y < 0$.
d) Using the yslices explain why $f_y(x, 0)$ does not exist for $x \neq 0$. What about $f_y(0, 0)$? Check the y slice though $(0, 0)$, and compute $f_y(0, 0)$ from difference quotients.

8. Let $f(x, y) = \sin(y - x^2)$. Apply the mfile `mslice` to this function at the point $P = (1, 1)$ as in Example 5.7.
a) In which direction $\mathbf{u}$ is $D_\mathbf{u} f(1, 1)$ largest? Does this agree with the direction of $\nabla f(1, 1)$?
b) In which directions $\mathbf{u}$ is $D_\mathbf{u} f(1, 1) = 0$? Is this direction tangent to the level curve at that point?

★ **9.** Write an mfile `u.m` for the function

$$u(x, y) = (-4x^3 + 3x^2 + 1)(y - y^2).$$

$u(x, y)$ is the temperature at a point (x, y) in the unit square $Q = \{0 \leq x, y \leq 1\}$. The heat flux at each point is the *negative* of the gradient vector, $-\nabla u(x, y) = -[u_x(x, y), u_y(x, y)]$.

(a) Verify by hand that $u_x(0, y) = 0$ and that $u = 0$ on the other edges of Q. This means that the left edge is insulated and that the temperature is held at zero on the other three edges.

(b) Put a mesh on Q with $\Delta x = \Delta y = .05$. Plot u on Q using the command `surf(X,Y,u(X,Y))`. Note where the highest temperature appears to be and the appearance of the surface on the edge $x = 0$.

(c) Compute the gradient of u by hand or symbolically. Write mfiles for u_x and u_y or define them as inline functions. Then enter the commands

```
>> U = u(X,Y);
>> Ux = ux(X,Y);
>> Uy = uy(X,Y);
>> contour(X,Y,U,20)
>> hold on
>> quiver(X,Y,-Ux,-Uy)
```

The last command puts arrows at each point of the meshgrid to represent the vector field of the heat flux, $(x, y) \rightarrow (-u_x(x, y), -u_y(x, y))$.

What is the direction of the heat flux with respect to the level curves? Where is the hot spot, and which way is the heat flowing?

What is the angle at which the level curves meet the edge $x = 0$? What is the direction of the heat flux at that edge of the square? How do you explain this physically?

Why are the flux vectors perpendicular to the other edges?

$\star$ **10.** Diffusion of a solute, such as salt, in a liquid medium is governed by Fick's law of diffusion. If $c(x, y, t)$ denotes the concentration of the solute at the point (x, y) and time t, the flux vector is $-k\nabla c = -k[c_x, c_y]$. k is the diffusion constant. As in the case of heat flow, the flux vector points in the opposite direction of the gradient, and the solute flows from areas of high concentration to low concentration. It can be shown that the concentration $c(x, y, t)$ satisfies the diffusion equation

$$c_t - k(c_{xx} + c_{yy}) = 0.$$

Let

$$c_1(x, y, t) = e^{-(5/4)t} \sin x \sin(y/2),$$

$$c_2(x, y, t) = e^{-(25/4)t} \sin(2x) \sin(3y/2),$$

and

$$c(x, y, t) = .2c_1(x, y, t) + c_2(x, y, t).$$

$c(x, y, t)$ is a solution of the diffusion equation, with $k = 1$. We shall observe the diffusion process on the square $Q = [0, \pi] \times [0, \pi]$. The concentration of the solute is assumed zero outside the square.

    a) Verify by hand that c solves the diffusion equation with $k = 1$.

    b) Make an mfile for c, c.m, as a function of three variables, x, y, t.

    c) We can view the concentration at a time t with the commands

```
[X,Y] = meshgrid(0:.1:pi);
C = c(X,Y,t);
pcolor(X,Y,C); shading flat; colorbar
hold on
contour(X,Y,C, 20,'k')
hold off
```

Using the subplot command, make a two-column display of the concentration at times $t = 0, .1, .2, .3, .4, .5$.

d) Through which sides of the square does the solute appear to be flowing? At each time, what is the point of greatest concentration?

11. Let $f(x, y) = x^2 + 3y^2$.

a) Graph f over the square $\{-2 \le x, y \le 2\}$ using qsurf.

b) Compute by hand the tangent plane approximation to f at the point $(x_1, y_1) = (-1, 1)$. Call this linear approximation $l_1(x, y)$. Do the same at the point $(x_2, y_2) = (1.5, .5)$. Call the approximation $l_2(x, y)$.

c) Use qsurf as in Example 5.10 to attach the tangent planes to the graph of f over the points (x_1, y_1) and (x_2, y_2). Graph the tangent planes over squares of side length 6.

$\star$ **12.** Let $f(x, y) = \exp(x) \cos(x + y)$.

a) Compute by hand the tangent plane approximation to f at the point $(-1, 1)$. Call this linear approximation $l(x, y)$.

b) Graph both f and l over the square $\{-2 \le x \le 0, 0 \le y \le 2\}$ using qsurf. Try to estimate $\max |E| = \max |f(x, y) - l(x, y)|$ over the square by examining the height between the graphs.

c) Now write an mfile that repeats the error estimating procedure of Example 5.10. Estimate the maximum error $f(x, y) - l(x, y)$ over squares $Q_h = \{|x - x_0|, |y - y_0| \le h\}$, where $(x_0, y_0) = (-1, 1)$ for $h = .5, .25, .125$. When h is halved, by what factor is the error reduced?

6

Functions of Three Variables and Parametric Surfaces

Prepared mfiles used in this chapter

```
arrow3 plane impl
```

Functions $f(x, y, z)$ of three variables can be defined as inline functions, in mfiles, and as symbolic expressions, just as for functions of one or two variables. The graph of a function of three variables is a three-dimensional surface in four-dimensional space, which we cannot visualize. One way to describe functions of three variables is in terms of their level surfaces. On the other hand, many surfaces in three-dimensional space are specified as level surfaces. We also discuss ways to display surfaces that are given parametrically.

6.1 Level sets and surfaces

When f is a function of three variables $f(x, y, z)$, the level sets

$$S_c = \{(x, y, z) : f(x, y, z) = c\}$$

usually consist of one or more two-dimensional surfaces called *level surfaces*. If we can graph the level sets of a function, we can get important information about its properties.

The level surfaces of quadratic functions of x, y, z are called *quadric surfaces*. Here are four important examples.

- A *paraboloid* is a level surface of the function
$$f(x, y, z) = z - x^2 - y^2.$$

  A level surface of f is
$$S_c = \{(x, y, z) : z - x^2 - y^2 = c\}$$

  and it is the graph of the function $z = g(x, y) = x^2 + y^2 + c$.

- A *hyperbolic paraboloid* is a level set of
$$f(x, y, z) = z - x^2 + y^2.$$

  A level surface here is the graph of $g(x, y) = x^2 - y^2 + c$ (several shown in Figure 6.1).

- A *hyperboloid* is a level surface of the function
$$f(x, y, z) = x^2 + y^2 - z^2.$$

  For $c \geq 0$, $S_c = \{f(x, y, z) = c\}$ is a *hyperboloid of one sheet*; for $c < 0$, it is a *hyperboloid of two sheets*. See Figure 6.2.

- An *ellipsoid* is a level surface of the function
$$f(x, y, z) = \frac{x^2}{a^2} + \frac{y^2}{b^2} + \frac{z^2}{c^2}.$$

  A special case is the sphere, when $a = b = c$.

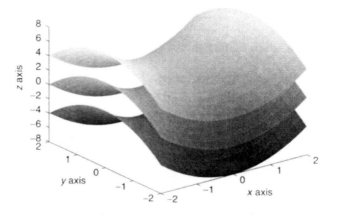

Figure 6.1 Hyperbolic paraboloids $z - x^2 + y^2 = c$ for $c = -4, 0, 4$.

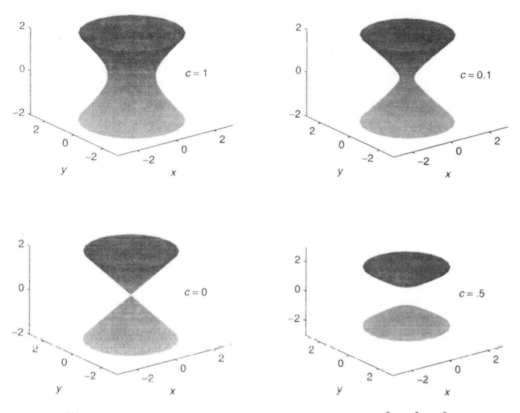

Figure 6.2 Hyperboloids, level surfaces of $f(x, y, z) = x^2 + y^2 - z^2$.

The level surfaces of these functions can all be graphed by solving for z and then using the usual commands for graphing functions $z = g(x, y)$. In some cases, such as the hyperboloids and the ellipsoids, the surfaces are not the graphs of a single function. For example, the sphere of radius $a > 0$ may be expressed as the union of the graphs of the two functions

$$g_+(x, y) = \sqrt{a^2 - x^2 - y^2}$$

$$g_-(x, y) = -\sqrt{a^2 - x^2 - y^2}$$

over the disk $\{x^2 + y^2 \le a^2\}$.

In the general case, we may not be able to solve for one of the variables in terms of the other two in the equation for the level set, $f(x, y, z) = c$. The following mfile constructs the level set by fixing a value of $z = z_0$ and uses the contour commands to find the level curves of $f(x, y, z_0) = c$ in the plane $z = z_0$. These level curves are then graphed together for a range of values of z_0 to generate the level set as a family of curves parallel to the x, y plane.

mfile impl.m   impl is short for implicit, because the level set is given implicitly by the equation $f(x, y, z) = c$. This mfile was written by Jonathan Rosenberg of the Department of Mathematics, University of Maryland.

The ingredients for the use of this mfile are a function $f(x, y, z)$ given in an mfile, or as an inline function, a set of coordinates defining a rectangular three-dimensional region, and a value c. The region is defined by a 6-vector corners, which is equal to $[x_{min}, x_{max}, y_{min}, y_{max}, z_{min}, z_{max}]$. The call is then impl(f,corners,c) when f is given an inline function and impl('f', corners,c) when f is given in an mfile.

Example 6.1

Consider $f(x, y, z) = x^2 + y^2 - z^2$, whose level sets are the hyperboloids shown in Figure 6.2. We show how the same surfaces can be constructed using impl. The results are shown in Figure 6.3.

```
f = inline('x.^2 + y.^2 - z.^2', 'x', 'y', 'z')
corners = [-2 2 -2 2 -2 2];
subplot(2,2,1)
  impl(f, corners, 1)
subplot(2,2,2)
  impl(f, corners, .1)
subplot(2,2,3)
  impl(f,corners, 0)
subplot(2,2,4)
  impl(f,corners, -.5)
```

Example 6.2

We take $f(x, y, z) = x^2 + y^2 + y^3 + z^2$ and display the level sets $f(x, y, z) = 1$ and $f(x, y, z) = .1$. See Figure 6.4.

```
f = inline('x.^2 + y.^2 + y.^3 + z.^2', 'x', 'y', 'z')
corners = [-2 2 -2 2 -2 2]
  subplot(1,2,1)
    impl(f, corners, 1)
  subplot(1,2,2)
    impl(f, corners, .1)
```

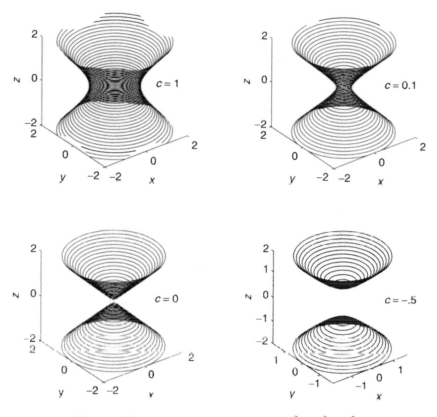

Figure 6.3 Level sets of $f(x, y, z) = x^2 + y^2 - z^2$.

Notice that in both Figures 6.3 and 6.4, for some values of c, the level sets can have more than one component. We shall see in Chapter 8 that level sets can be useful in determining maxima and minima of functions.

6.2 Color slices of a solid

A second way to gain information about a function $f(x, y, z)$, which complements the graphing of level sets, is to slice through its domain with planes parallel to the coordinate axes and show the values of f on these planes with color. This is very helpful when it is difficult to solve for the level sets. This technique can also be used to make rough guesses of the location of local maxima and minima.

To illustrate these ideas we imagine a solid material heated to a temperature that depends on the location. The temperature is a function $f(x, y, z)$. Suppose we can tell the temperature of the material at a given point by observing the color of the

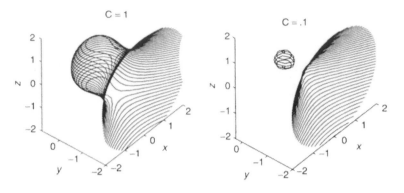

Figure 6.4 Level sets of $f(x, y, z) = x^2 + y^2 + y^3 + z^2$.

material according to the following scale:

temperature	color
$0°$	black
$20°$	brown
$40°$	red
$60°$	orange
$80°$	yellow
$100°$	white

Then to get an idea of the temperature distribution within the material, we can take slices of the material in various directions and observe the colors. Level sets of the temperature function will correspond to surfaces of constant color.

The MATLAB command `slice` implements this idea. Since we are dealing with functions of three variables, we must use three-dimensional arrays of coordinates.

Example 6.3

Let R be the three-dimensional region

$$R = \{(x, y, z) : -2 \leq x, y \leq 2, \ 0 \leq z \leq 4\}$$

and

$$f(x, y, z) = 20(z + e^{-x^2 - y^2}).$$

We put a three-dimensional array of meshpoints in R as follows. We pick x coordinate values $x_i = -2, -1.8, \ldots, 1.8, 2$, y coordinate values, $y_j = -2, -1.8, \ldots,$ 1.8, 2 and z coordinate values, $z_k = 0, .2, .4, \ldots, 3.8, 4$. Then the coordinate mesh consists of the three-dimensional array of triples (x_i, y_j, z_k). The function values at these meshpoints constitute another three-dimensional array, $w_{i,j,k} = f(x_i, y_j, z_k)$. These arrays are constructed with the following MATLAB commands:

```
f =
inline('20*(z + exp(-x.^2 - y.^2))', 'x', 'y', 'z')
x = -2:.2:2; y = x;
z = 0:.2:4;
[X,Y,Z] = meshgrid(x,y,z);
W = f(X,Y,Z)
```

Each of the arrays X, Y, Z, W has dimensions $21 \times 21 \times 21$. We can now slice through the solid region R with planes parallel to the coordinate planes and observe the colors of the function on each of these planes. For example, we might choose the planes $x = 0$, $y = 0$, and $z = 2$. The commands to produce these planes and display them are

```
slice(X,Y,Z,W,0,0,2)
colormap(hot); colorbar
```

The result is shown in Figure 6.5. On the page here, there is no color, just shades of gray. However, on the screen of your computer you will see the colors range from black to white, through red and yellow, using the colormap hot.

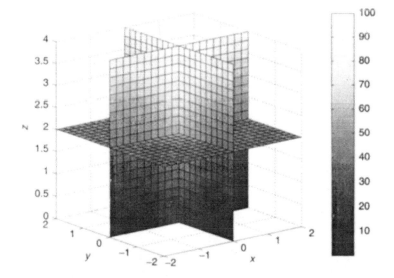

Figure 6.5 Slices through the solid region R, with colors determined by the function $f(x, y, z) = 20(z + \exp(-x^2 - y^2))$.

We can see that the temperature rises as z increases, and reaches its greatest value of 100 at the point $(0, 0, 4)$. This is shown by the colorbar that makes the correspondence between color and the numerical values, $0 \leq z \leq 100$, in Figure 6.5.

6.3 The gradient vector field

The gradient vector of a function of three variables is

$$\nabla f(x, y, z) = [f_x(x, y, z), f_y(x, y, z), f_z(x, y, z)].$$

Just as for functions of two variables, the gradient vector at a point $P_0 = (x_0, y_0, z_0)$ is orthogonal to the level surface S of f through P_0. By this we mean that if $\mathbf{r}(t)$ is any smooth curve on the surface S with $\mathbf{r}(0) = P_0$, then the tangent vector to the curve at $t = 0$, which is $\mathbf{r}'(0)$, is orthogonal to $\nabla f(x_0, y_0, z_0)$.

The gradient vector field can be displayed using the MATLAB command quiver3, which is the three-dimensional analog of quiver that we used in Chapter 5. The call is quiver3(X,Y,Z,U,V,W). Here X,Y,Z are matrices that describe some surface in three-dimensional space. For instance, we could have Z = g(X,Y) for some function g. U,V,W are the matrices of the components of vectors that are to be attached at each of the points (x, y, z) that lie on the surface.

Example 6.4

Let $f(x, y, z) = z + (y^2 - x^2)/4$. The level surfaces of f are hyperbolic paraboloids. The gradient vector field is $\nabla f(x, y, z) = [-x/2, y/2, 1]$. We display the gradient vector field in two ways. First we attach gradient vectors to points in a family of planes parallel to the x, y plane at heights $z = -1, 0, 1$ over the square $0 \leq x, y \leq 2$. This display is shown on the left of Figure 6.6.

```
[X,Y] = meshgrid(0:.4 :2);
U = -X/2;
V = Y/2;
W = 1+0*X;
subplot(1,2,1)
for z = [-1,0,1]
   Z = z +0*X;
   quiver3(X,Y,Z,U,V,W)
   hold on
end
axis image
```

Only three z levels were chosen; too many levels produce a very confusing figure. In versions 5.3 and higher, quiver3 can also take three-dimensional arrays as arguments. See Section 11.1.

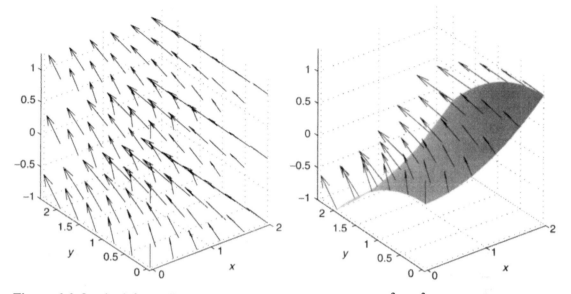

Figure 6.6 On the left, gradient vector field of $f(x, y, z) = z + (y^2 - x^2)/4$. On the right, level surface $f(x, y, z) = 0$, and gradient vector field.

Our second display of these gradient vectors attaches them to points on the portion of the level surface $f(x, y, z) = 0$ lying over the same square, $0 \le x, y \le 2$. The level surface $f = 0$ has the equation $z = (x^2 - y^2)/4$. We add the following commands to the preceding script:

```
% First plot the surface with finer mesh.
[XX,YY] = meshgrid(0: .05: 2);
ZZ = .25*(XX.^2 - YY.^2);
subplot(1,2,2)
surf(XX,YY,ZZ); shading interp
hold on

% Now add the gradient vector field.
Z = .25*(X.^2 - Y.^2);
quiver3(X,Y,Z,U,V,W)
axis image
```

This second display is shown in the right side of Figure 6.6.

The tangent plane to a surface S at the point $P_0 \in S$ is the collection of all vectors attached at P_0 and tangent to S at P_0. In other words, the tangent plane is the set of all vectors attached at P_0 that are orthogonal to a normal vector to S at that point. In the case of a level surface

$$S = \{(x, y, z) : f(x, y, z) = c\},$$

a normal direction to the surface is given by

$$\nabla f = [f_x(P_0), f_y(P_0), f_z(P_0)].$$

Hence the equation of the tangent plane is given by

$$f_x(P_0)(x - x_0) + f_y(P_0)(y - y_0) + f_z(P_0)(z - z_0) = 0. \tag{6.1}$$

Of course this equation defines a plane only if at least one of the components of $\nabla f(P_0)$ is nonzero. Note that in the case of the hyperboloid $f(x, y, z) = x^2 + y^2 - z^2 = 0$ shown in Figure 6.2, the point $P_0 = (0, 0, 0)$ lies on the surface at the vertex of the double cones, and $\nabla f(P_0) = 0$. The surface is singular at this point and there is no tangent plane defined.

6.4   Parametric representation of surfaces

When surfaces are not expressible as the graph of a function $z = g(x, y)$, we often use a *parametric* representation. Intuitively we think of starting with a flat piece of surface, usually a rectangle or a disk, in the plane and then curving it in space to form the desired surface. We use coordinates (u, v) to locate points in the initial piece of surface D. The curving in space is the same as defining three functions on D for the three spatial coordinates,

$$x(u, v), \quad y(u, v), \quad z(u, v).$$

The mapping $(u, v) \rightarrow (x(u, v), y(u, v), z(u, v)) : D \rightarrow R^3$ is the parametric representation of the surface S that is the image of this mapping.

Recall the parametric representation of a plane that we used in Chapter 3. Here we take D to be all of R^2. Suppose the plane passes through the point $P_0 = (x_0, y_0, z_0)$ and has normal vector $\mathbf{N}$. Let $\mathbf{v}$ and $\mathbf{w}$ be two vectors orthogonal to $\mathbf{N}$. Then all points in the plane can be represented $P = P_0 + s\mathbf{v} + t\mathbf{w}$. Here we use s, t instead of u, v as the parameters. The coordinate functions are

$$x(s, t) = x_0 + sv_1 + tw_1, \qquad y(s, t) = y_0 + sv_2 + tw_2, \qquad z(s, t) = z_0 + sv_3 + tw_3.$$

If we want to represent a piece of the plane containing P_0, we restrict the parameters to lie in a finite set D, say, $D = \{-1 \leq s, t \leq 1\}$.

Example 6.5

We wish to find a parametric representation of the cylinder of radius $a > 0$ that is centered about the z axis and that extends from $z = -\pi/2$ to $z = \pi/2$. Now we start with a domain D, which is a rectangle, $D = \{0 \leq u \leq 2\pi, \ -\pi/2 \leq v \leq \pi/2\}$. We want to roll up this flat piece of surface to form the cylinder. We shall do so by making the line segments parallel to the u axis into circles of radius a and the line segments parallel to the v axis into the vertical lines on the surface of the cylinder. A set of coordinate functions that will do this is

$$x(u, v) = a\cos(u), \qquad y(u, v) = a\sin(u), \qquad z(u, v) = v.$$

To graph this surface with MATLAB we proceed as follows, taking $a = 1$:

```
% Define coordinates in u,v space
u = linspace(0, 2*pi, 21);
v = linspace(-pi/2, pi/2, 21);
[U,V] = meshgrid(u,v);
a = 1;

% Define the coordinate functions
X = a*cos(U);
Y = a*sin(U);
Z = V;
surf(X,Y,Z)
```

The result is shown in Figure 6.7.

Example 6.6

Next we tackle the problem of parameterizing the sphere of radius a. We continue the process we started in the previous example, where we parameterized a cylinder.

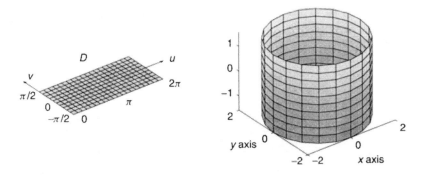

Figure 6.7 Flat piece D of the u, v plane rolled into cylinder in x, y, z space.

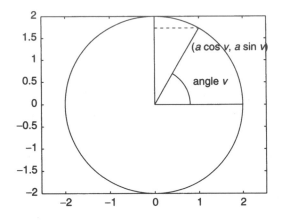

Figure 6.8 Slice through sphere of radius a.

To make a sphere, we want to "pinch in" the top and bottom of the cylinder. The vertical lines on the sides of the cylinder will become the meridians on the sphere; the circle in the x, y plane will become the equator on the sphere. Considering Figure 6.8, we see that the following coordinate functions accomplish this:

$$x(u, v) = a\cos(v)\cos(u), \qquad y(u, v) = a\cos(v)\sin(u), \qquad z(u, v) = a\sin(v),$$

$$0 \le u \le 2\pi, \quad -\pi/2 \le v \le \pi/2.$$

u is longitude and v is latitude on the sphere. In this parameterization, the north pole corresponds to $v = \pi/2$ and the south pole to $v = -\pi/2$. In Figure 6.9, we see the sphere produced by the following commands and the same sphere with a patch that is the image of $0 \le u \le \pi/2$, $0 \le v \le \pi/3$:

```
a = 2;
u = linspace(0, 2*pi, 41);
```

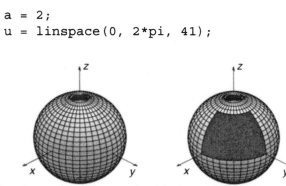

Figure 6.9 On the left, sphere of radius $a = 2$ with lines of latitude and longitude. On the right, patch on sphere is the image of $0 \le u \le \pi/2$, $0 \le v \le \pi/3$.

```
v = linspace(-pi/2, pi/2, 31);
[U,V] = meshgrid(u,v);
X = a*cos(V).*cos(U);
Y = a*cos(V).*sin(U);
Z = a*sin(V);
surf(X,Y,Z); colormap(gray); shading flat
```

The parameterization of the sphere just given is not the only one. The parameterization of the sphere of radius r used in spherical coordinates is

$$x(r, \theta, \phi) = r \sin(\phi) \cos(\theta), \quad y(r, \theta, \phi) = r \sin(\phi) \sin(\theta), \quad z(r, \theta, \phi) = r \cos(\phi).$$

Here the azimuthal angle is θ, $0 \le \theta \le 2\pi$, and ϕ is the polar angle, $0 \le \phi \le \pi$. In spherical coordinates, the north pole corresponds to $\phi = 0$ and the south pole to $\phi = \pi$.

Finally, if you want to quickly graph a unit sphere, the single MATLAB command `sphere` will display a unit sphere. For more information, enter `help sphere`.

Surfaces of revolution

The sphere and the cylinder both fall into the class of surfaces called *surfaces of revolution*. They can easily be represented parametrically. We shall discuss surfaces of revolution that are generated by revolving a curve C in the x, z plane about the z axis.

Let the curve C in the x, z plane be parameterized by $v \rightarrow (x_0(v), z_0(v))$. The y coordinate $y_0(v) \equiv 0$. We want to rotate this curve around the z axis, leaving the z coordinate unchanged.

Recall from Chapter 4 that the rotation matrix $\mathbf{R}$ that rotates points in the x, y plane through an angle u is

$$\mathbf{R} = \begin{bmatrix} \cos(u) & -\sin(u) \\ \sin(u) & \cos(u) \end{bmatrix}.$$

The x, y coordinates of the rotated curve will be given by

$$\begin{bmatrix} x(u, v) \\ y(u, v) \end{bmatrix} = \begin{bmatrix} \cos(u) & -\sin(u) \\ \sin(u) & \cos(u) \end{bmatrix} \begin{bmatrix} x_0(v) \\ y_0(v) \end{bmatrix}.$$

Since $y_0(v) = 0$ we find

$$
\begin{aligned}
x(u, v) &= \cos(u)x_0(v) \\
y(u, v) &= \sin(u)x_0(v) \\
z(u, v) &= z_0(v).
\end{aligned}
\tag{6.2}
$$

Example 6.7

The torus is the surface of a doughnut. We could continue our method of describing the surfaces in terms of a rolling and bending of the rectangular flat domain D in the u, v plane. In this case, we could think of rolling the domain D into a tube and then bending the tube around until it formed the donut.

Instead, we shall view the torus as a surface of revolution. Consider the circle of radius a in the x, z plane, with center at $(r, 0, 0)$, where $r > a$. The torus is the surface swept out by the circle as the plane of the circle is revolved about the z axis. See Figure 6.10.

Now, the circle in the x, z plane is parameterized by

$$x_0(v) = r + a\cos(v), \qquad y_0(v) = 0, \qquad z_0(v) = a\sin(v), \qquad 0 \le v \le 2\pi.$$

The parameterization (6.2) becomes

$$x(u, v) = \cos(u)x_0(v) = \cos(u)(r + a\cos(v))$$
$$y(u, v) = \sin(u)x_0(v) = \sin(u)(r + a\cos(v))$$
$$z(u, v) = a\sin(v).$$

The torus is graphed using the following commands. See Figure 6.11.

```
a = .5; r = 2;
u = linspace(0, 2*pi, 41); v = u;
[U,V] = meshgrid(u,v);
X = cos(U).*(r + a*cos(V));
Y = sin(U).*(r + a*cos(V));
Z = a*sin(V);
surf(X,Y,Z);
```

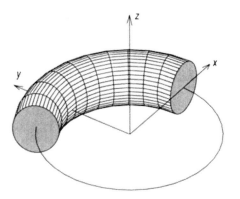

Figure 6.10 Circle rotating about z axis to generate torus.

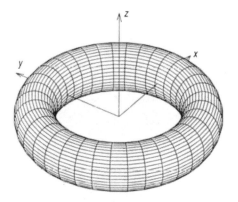

Figure 6.11 Torus with $r = 2$ and $a = .5$.

MATLAB has a single command to graph a surface of revolution when the curve C in the x, z plane is given as a function $x = f(z)$, $0 \leq z \leq 1$. For example, if $x = f(z) = 3(z - 1/3)^2$, the resulting surface of revolution is produced by

```
z = linspace(0,1,41);
cylinder(3*(z-1/3).^2)
```

See Figure 6.12.

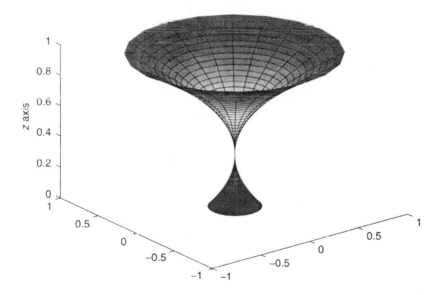

Figure 6.12 Surface of revolution with generating curve $x = 3(z - 1/3)^2$.

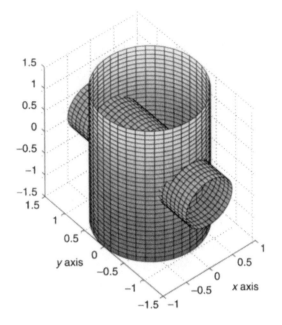

Figure 6.13 Vertical cylinder of radius 1 and horizontal cylinder of radius .5.

Of course, it is possible to combine graphs to depict more complicated objects. Example 6.8 shows how two intersecting cylinders can be displayed, see Figure 6.13.

Example 6.8

```
u = linspace(0,2*pi,41);
v = linspace(-2,2,41)
[U,V] = meshgrid(u,v);

% vertical cylinder of radius 1
surf(cos(U), sin(U), V);
hold on

% horizontal cylinder of radius .5
surf(.5*cos(U), V, .5*sin(U))
hold off
```

The ezsurf command

The command `ezsurf` can also be used to graph surfaces given parametrically in terms of symbolic expressions. The syntax is `ezsurf(x,y,z,[a b c d])`. The parameters are s and t, with $a \leq s \leq b$ and $c \leq t \leq d$. x, y, z are symbolic expressions given in terms of s and t.

Example 6.9

Here is another example of two surfaces glued together, this time to form a coffee (or beer) mug. See Figure 6.14.

```
syms s t
% vertical cylinder of radius 1
x = cos(s);
y = sin(s);
z = t;
ezsurf(x,y,z, [0 2*pi -2 2])
hold on

% handle formed by half of a torus with r = 1, a = .25
% centered at (1,0,.5)
xhandle =  1 + cos(s)*(1+.25*cos(t));
yhandle = .25*sin(t);
zhandle = .5 + sin(s)*(1+.25*cos(t));
ezsurf(xhandle, yhandle, zhandle, [ -pi/2 pi/2 0 2*pi])
hold off
axis([-2 3 -2 2 -2 2])
```

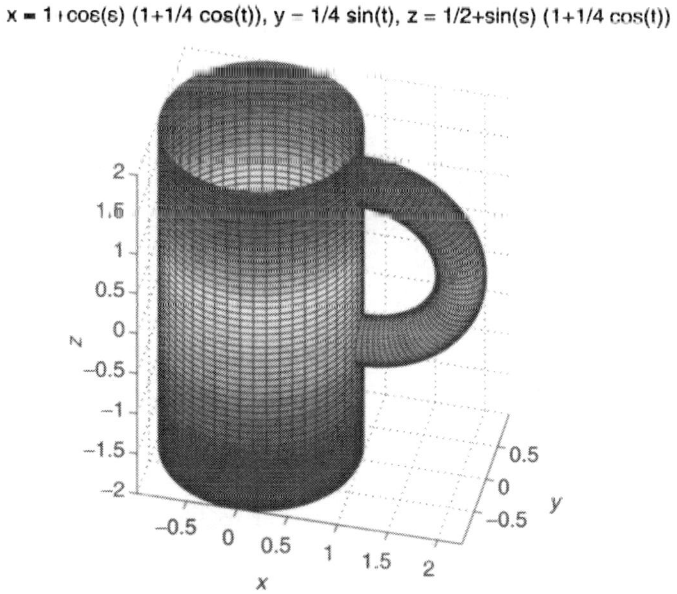

Figure 6.14 Mug, generated by symbolic expressions and `ezsurf`.

6.5   Normal vectors and tangent planes in parametric form

In Figures 6.7 and 6.9–6.13, the curves on each surface are the images of the grid lines in the u, v plane. For example, in Figure 6.9, the curves where v is constant are the circles of constant latitude, and the curves where u is constant are the meridians of constant longitude.

More generally, when S is given parametrically by

$$(u, v) \to (x(u, v), \ y(u, v), \ z(u, v))$$

a tangent vector to the curve $u \to (x(u, v_0), \ y(u, v_0), \ z(u, v_0))$ is

$$\mathbf{v} = [x_u(u, v_0), \ y_u(u, v_0), \ z_u(u, v_0)].$$

A tangent vector to the curve $v \to (x(u_0, v), y(u_0, v), z(u_0, v))$ is

$$\mathbf{w} = [x_v(u_0, v), \ y_v(u_0, v), \ z_v(u_0, v)].$$

These two vectors span the tangent plane at the point $(x_0, y_0, z_0) = (x(u_0, v_0), y(u_0, v_0), z(u_0, v_0))$. Hence a direction normal to the tangent plane at this point is

$$\mathbf{n} = \mathbf{v} \times \mathbf{w}.$$

We define a *normal direction to the surface* at the point (x_0, y_0, z_0) to be this vector $\mathbf{n}$. In terms of the coordinate functions this normal vector can be written

$$\mathbf{n}(u, v) = [y_u z_v - y_v z_u, \ x_v z_u - x_u z_v, \ x_u y_v - x_v y_u]. \tag{6.3}$$

Usually we shorten our terms and say simply that Eq. (6.3) is *the* normal to the surface, although any multiple of $\mathbf{n}$ is also a normal direction.

In the case of a surface of revolution, using Eqs. (6.2) and (6.3), we have

$$\mathbf{n}(u, v) = [\cos u \ x_0(v) z_0'(v), \ \sin u \ x_0(v) z_0'(v), \ -x_0(v) x_0'(v)]. \tag{6.4}$$

For example, the sphere of radius a is a surface of revolution, and we recall the parameterization for $0 \le u \le 2\pi$, $-\pi/2 \le v \le \pi/2$:

$$x(u, v) = a \cos(u) \cos(v), \qquad y(u, v) = a \sin(u) \cos(v), \qquad z(u, v) = a \sin(v).$$

Hence, substituting in Eq. (6.4), the normal to the sphere is given by

$$\begin{aligned}
\mathbf{n}(u, v) &= a^2 [\cos u \cos^2(v), \ \sin u \cos^2(v), \ \cos v \sin v] \\
&= a \cos v \ [x(u, v), \ y(u, v), \ z(u, v)].
\end{aligned}$$

Exercises

1. Let $f(x, y, z) = x^2 + y^2 + z^2 + xy^2$.

a) Use the mfile `impl` over the region $\{-4 \le x, y \le 1, -2 \le z \le 2\}$. Display the level sets $S_c = \{f = c\}$ for $1.2, 1, .8, .5, .3, .2$. Plot these graphs in the same figure window using the command `subplot`, as in Example 6.1. Off the surface S_c, the function f is not equal to c. On each graph indicate on which side of S_c we have $f < c$ and $f > c$.

b) For what value of c does the level set S_c break into two components?

c) As c gets smaller, the component of S_c that contains the origin resembles a sphere. Can you explain this by looking at the formula for f?

2. Let
$$f(x, y, z) = e^{-(x-1)^2 - y^2 - z^2} + e^{-(x+1)^2 - y^2 - z^2}.$$

a) Use `impl` over the region $-2 \le x, y, z \le 2$. Display the level set S_c for $c = .2, .5, .7, .8$. Plot them in the same figure, as in Example 6.1.

b) By experimenting with various values of c, find that value c_* such that the level set S_c has only one component for $c < c_*$ and has two components for $c > c_*$.

c) The level set S_{c_*} will have a cusp at $(0, 0, 0)$. There is no tangent plane to the surface defined there. Verify that $\nabla f(0, 0, 0) = 0$.

3. Suppose seismic data tells us that the density of rock in some region is given by

$$\rho(x, y, z) = (1 - z)e^{-.2x - .3y^2} + 101.$$

We want to study the density function in the region $R = \{0 \le x, y, z \le 2\}$. Make a three-dimensional grid on R, with 40 subdivisions in each direction. Use the command `slice` to make a set of slices through $(0, 0, 0)$. Use `rotate3d` to rotate the figure so that you view the region from the point $(2, 2, 2)$. Add the colorbar.

a) Where is the rock density greatest?

b) The rock density is nearly constant along the line $x = 2, z = 0$. Using the colorbar, estimate the constant density along this line.

c) Graph the function $y \to \rho(2, y, 0)$ to compare with the estimate from the `slice` portrait.

4. Let the region $R = \{0 \le x, y, z \le \pi\}$ and let the temperature be given by $u(x, y, z) = \cos x \sin(2y) \sin(3z/2)$.

a) Make a three-dimensional grid on R, with 40 subdivisions in each direction. Use `slice` with various collections of planes and the colorbar with `colormap(hot)`. Determine the hottest and coldest points in the region.

b) The heat flux $F(x, y, z) = -\nabla u(x, y, z)$. Use the commmand `quiver3` to plot the heat flux vector on the three planes $z = 0, z = \pi/2$, and $z = \pi$. You will need to write three mfiles or inline functions for the three derivatives $-u_x, -u_y$, and $-u_z$. On which sides of the cube is the flux vector tangent to the side? These are the insulated sides through which no heat flows. On which sides of the cube is the flux vector orthogonal to the side?

5. The hyperboloids of Figure 6.2 are surfaces of revolution. The section of the hyperboloid in the x, z plane satisfies $x^2 - z^2 = c$.

a) For $c > 0$ and $x > 0$, verify by hand that we can parameterize the curve by

$$x_0(v) = \sqrt{c}\cosh(v), \qquad z(v) = \sqrt{c}\sinh(v).$$

b) Following Example 6.7, make a script file to graph the hyperboloids for several values of $c > 0$.

c) Find the parameterization for $c < 0$, and graph the hyperboloids for several values of $c < 0$. In this case there are two sheets, and you will need two different expressions for $z(v)$.

6. Let $x_0(v) > 0$ and $z(v)$ parameterize a curve in the x, z plane.

a) What is the parameterization of the surface of revolution obtained by revolving this curve about the z axis?

b) Modify your script file of Exercise 5 to graph this surface of revolution for the following choices of $x_0(v)$ and $z(v)$:
  (i) $x_0(v) = \exp(-v^2/2)$, $z(v) = v$, $-2 \le v \le 2$.
  (ii) $x_0(v) = \exp(-v^2/10)\cos^2(v) + 1$, $z(v) = v$, $-2\pi \le v \le 2\pi$.
  (iii) $x_0(v) = |\sin(v)| + 1$, $z(v) = v$, $-3\pi \le v \le 3\pi$.

7. a) Parameterize the section in the x, z plane of the ellipsoid

$$x^2 + y^2 + z^2/9 = 1.$$

b) Graph it as a surface of revolution (revolved about the z axis).

★   **8.** Let $f(u) = \cos(u)$, $0 \le u \le \pi$. Let $g_{top}(u) = .05\sin(u)\exp(-2\cos(u))$, and $g_{bottom}(u) = -g_{top}(u)/2$, again for $0 \le u \le \pi$.

a) Plot both curves $u \to (f(u), g_{top}(u))$ and $u \to (f(u), g_{bottom}(u))$ on the same graph. Then use `axis equal`. The resulting graph is the cross section of a wing.

b) Now make a three-dimensional graph of the wing with this cross section. The wing should lie along the y axis, extending for $0 \le y \le 4$. You will need to make the graph in two pieces, top and bottom.

c) Use the MATLAB numerical integrator `quadl` to find the arc length of the curve bounding the cross section. Then find the total surface area of the wing.

★ **9.** Write a function mfile, `bar.m`, that graphs a bar of radius d with axis of symmetry an arbitrary line through the origin. Let $\mathbf{L} = [a, b, c]$ be a unit tangent vector to the line of symmetry. The call for the function mfile should be `bar(a,b,c,d)`.

a) First find a formula for a pair of unit vectors $\mathbf{u}$ and $\mathbf{v}$ such that $\mathbf{u} \cdot \mathbf{v} = \mathbf{u} \cdot \mathbf{L} = \mathbf{v} \cdot \mathbf{L} = 0$.

b) Then parametcrize the cylindrical surface of the bar in terms of s and t:

$$(x, y, z) = t\mathbf{L} + (d \cos s)\mathbf{u} + (d \sin s)\mathbf{v}.$$

What is the range of the parameter s? You can choose any range of the parameter t that will produce a good graph.

c) Now make the formula more flexible to allow the radius d to depend on t.

★ **10.** Let $x(t) = r \cos(t)$, $y(t) = r \sin(t)$, and $z(t) = at$ parameterize a circular helix of radius $r > 0$ with rise coefficient $a \geq 0$.

a) Use parameters s and t to parameterize the surface of the tube of radius b that follows the helix. The circular cross section of the tube should lie in the plane spanned by the vectors $-\mathbf{N} = [\cos t, \sin t, 0]$ and $\mathbf{B} = [a \sin t, -a \cos t, r]/\sqrt{r^2 + a^2}$ of the Frenet frame.

b) Write a function mfile, `tube.m`, with call `tube(r,a,b)`, that graphs the tube for $0 \leq t \leq 4\pi$.

★ **11.** Extend the method of Exercise 10 to parameterize a tube following the helix of Exercise 10 with different cross sections.

a) Elliptical cross sections

b) Circular cross sections with radius b that depend on t.

★ **12.** Use symbolic expressions and the command `ezsurf` as in Example 6.9 to graph a cylindrical bar of radius $1/4$ running from $(-1, -1, -1)$ to $(1, 1, 1)$. Then add two spheres of radius 1, one centered at each end of the bar.

13. Recall the parameterization of the torus,

$$
\begin{aligned}
x(u, v) &= \cos(u)(r + a \cos(v)) \\
y(u, v) &= \sin(u)(r + a \cos(v)) \\
z(u, v) &= a \sin(v), \qquad 0 \leq u, v \leq 2\pi.
\end{aligned}
$$

a) Find an expression for the normal $\mathbf{N}(u, v)$ on the surface.

b) Graph the torus and attach the tangent plane at several points. Use the mfile `plane`.

14. A *ruled surface* is one that is swept out by a straight line moving along a curve. We parameterize the curve by $\mathbf{r}(t) = [x_0(t),\ y_0(t),\ z_0(t)]$. Let the tangent vector to the line at each t be $\mathbf{L}(t) = [a(t),\ b(t),\ c(t)]$. Then the ruled surface is parameterized by

$$\mathbf{r}(t) + s\mathbf{L}(t) = [x_0(t) + sa(t),\ y_0(t) + sb(t),\ z_0(t) + sc(t)].$$

a) Let $\mathbf{r}(t) = (x_0(t), y_0(t), z_0(t)) = (2\cos(t), 2\sin(t), 0)$ parameterize the circle of radius 2 in the x, y plane. Let $\mathbf{L}(t) = [\cos(t/2),\ \sin(t/2),\ \sin(t/2)]$. Graph the surface for $0 \le t \le 2\pi$ and $-.2 \le s \le .2$. The resulting surface is called the *Moebius* band.

b) Calculate the normal vector $\mathbf{N}$ to the surface at points $t = 0, \pi/6, \pi/3,$ $\ldots, 5\pi/6$ and $s = 0$. Use the mfile `arrow3` to attach the normal vector to the graph of the surface at these points. What is the limiting position of the normal as t approaches 2π? Is it possible to construct normal vectors on the surface that are globally continuous?

★★   **15.** Graph a crystal structure with spheres of small radius at the origin and at each of the eight corners of the cube $-1 \le x, y, z \le 1$. Connect each of the spheres with slender straight tubes to represent the bonds.

★★   **16.** Graph a door with a door knob.

★★   **17.** Graph a bolt. The thread on the bolt can be drawn with several helixes. The smaller radius of the bolt, on the inside of the thread, should be $r = 1$, and the larger radius should be $r = 1.1$. The pitch of the thread is the angle the helix makes with the horizontal. Make the pitch 10 degrees.

7

Solving Equations

Prepared mfiles used in this chapter

`newton2`

In this chapter we shall be concerned mainly with solving a system of two equations in two unknowns,

$$f(x, y) = 0 \qquad (7.1)$$
$$g(x, y) = 0.$$

In the first section we show how to solve a system of this type using the symbolic `solve` command that we saw in Section 1.6. We then discuss numerical methods to solve a single equation, $f(x, y) = 0$, and a system, Eqs. (7.1). The numerical methods provide very important applications of the tangent line and tangent plane approximations of a function.

7.1 Symbolic solutions

We illustrate how to use the symbolic solver with an example.

Example 7.1

Let $f(x, y) = y - 4x^2 + 3$ and $g(x, y) = x^2/4 + y^2 - 1$. The graphs of the level curves $f(x, y) = 0$ and $g(x, y) = 0$ are shown in Figure 7.1. We see that there are four roots of system (7.1), at the points A, B, C, D where the level curves $f = 0$ intersect the level curves $g = 0$.

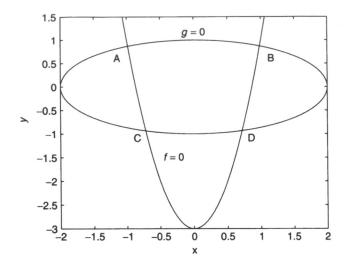

Figure 7.1 Zero curves of $f(x, y) = y - 4x^2 + 3$ and $g(x, y) = x^2/4 + y^2 - 1$.

We define symbolic functions f and g of x and y. Then we apply the `solve` command with two arguments. The command assumes that we want to solve Eqs. (7.1).

```
syms x y
f = y - 4*x^2 +3;
g = .25*x^2 + y^2 -1;
[a b] = solve(f,g);
[a b]
ans =

[   1/16*(190+14*17^(1/2))^(1/2),   -1/32+7/32*17^(1/2)]
[  -1/16*(190+14*17^(1/2))^(1/2),   -1/32+7/32*17^(1/2)]
[   1/16*(190-14*17^(1/2))^(1/2),   -1/32-7/32*17^(1/2)]
[  -1/16*(190-14*17^(1/2))^(1/2),   -1/32-7/32*17^(1/2)]

double([a b])
ans =

   0.9837     0.8707
  -0.9837     0.8707
   0.7188    -0.9332
  -0.7188    -0.9332
```

When the symbolic solver is "stumped," it will try to find at least one solution numerically.

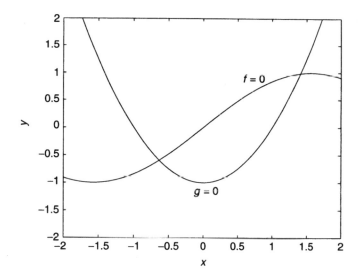

Figure 7.2 Zero-level curves of $f(x, y) = y - \sin(x)$ and $g(x, y) = y - x^2 + 1$.

Example 7.2

Consider $f(x, y) = y - \sin(x)$ and $g(x, y) = y - x^2 + 1$. The zero-level curves are shown in Figure 7.2. There are solutions near $(x, y) = (-.5, -.5)$ and $(x, y) = (1.5, 1)$.

```
syms x y
f = y - sin(x);
g = y - x^2 +1;
[a b] = solve(f,g);
[a b]
ans =
[ 1.4096240040025962492355939705895,
          .98703983266031148648572656479133]
```

The equation solver has found only one of the roots, and has found it numerically. We do not have much control over which roots it will find. Thus it is important to understand how to use numerical methods to solve systems like Eqs. (7.1).

7.2 Numerical solutions in one dimension

The most well-known numerical algorithm for solving equations

$$f(x) = 0 \tag{7.2}$$

is Newton's method, which is discussed in every calculus book. Recall the formula for the iterations,

$$x_{n+1} = x_n - f(x_n)/f'(x_n). \tag{7.3}$$

If the root x_* is a *simple root* ($f'(x_*) \neq 0$), Newton's method converges rapidly when a good starting value x_1 is provided. A poor starting value can cause the method to converge to the wrong root or even to fail to converge.

Newton's method can also be used to find a root x_* of $f(x) = 0$, where $f'(x_*) = 0$, but the convergence is much slower.

Here is a simple script mfile that implements Newton's method. In addition you will need to write mfiles f.m and df.m for the function and its derivative. Input at runtime is the starting value x_1 and the number of iterations.

```
xstart = input('enter the starting value    ')
N = input(' enter the number of iterations desired   ')
% We set aside N memory spaces for the iterates.
x = zeros(N,1);
x(1) = xstart;
for n = 1:N-1
    x(n+1) = x(n) - f(x(n))/df(x(n));
end
[x, f(x)]
```

This version displays all the iterates x_n and the function values $f(x_n)$.

It is very important to know how close the approximation x_n is to the exact root. Since we do not know the exact root x_* (otherwise we would not need a root finder), we must be able to estimate the error, $x_n - x_*$, using only the computed values. It can be shown that when x_* is a simple root of $f(x) = 0$, we have the practical estimate (not an error bound)

$$|x_n - x_*| \approx |x_n - x_{n-1}|.$$

Thus if we required our approximate root to be within 10^{-4} of the exact root, we would choose N so large that $|x_N - x_{N-1}| \leq 10^{-4}$, and then we would take x_N as our estimate of the root. Later, in Chapter 13, we shall see how to incorporate this choice of the number of iterations in the code with a while statement.

Example 7.3

Let $f(x) = \sin(x) - x/2$. From Figure 7.3, we see that in addition to the root $x = 0$, there are roots near $x = \pm 2$. With a starting value of $x_1 = 1.2$, the method converges rapidly to the root near $x = 2$.

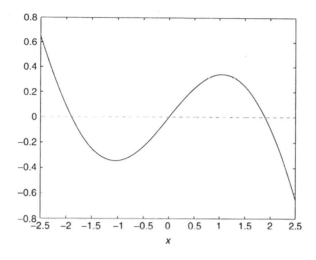

Figure 7.3 Graph of $f(x) = \sin(x) - x/2$.

iterate	x	f(x)
1	1.20000000000000	0.33203908596723
2	3.61233412536092	-2.25971429622269
3	1.98808029480705	-0.07984708586772
4	1.89987865948990	-0.00360002090311
5	1.89550532205812	-0.00000905437169
6	1.89549426710469	-0.00000000005791

However, if the starting value is $x_1 = 1.1$, the method does not converge. The first ten iterates are

iterate	x	f(x)
1	1.100000000000	.341207360061
2	8.452992261503	-3.400602278156
3	5.256413636975	-3.483839338117
4	203.384183718982	-100.961397298552
5	118.019330723405	-59.987778289229
6	-87.470869259029	44.209343971404
7	203.635887828739	101.280304759961
8	-128.232234454577	63.573907843189
9	-80.797636016021	41.171955352153
10	-387.504830399606	194.638548604918

The first guess at the root, $x_1 = 1.1$, is too close to the maximum of the function. The line tangent to the graph at $x_1 = 1.1$ is too close to horizontal and intersects the x axis far to the right of the root.

The root finder fzero

The MATLAB numerical root finder `fzero` was mentioned in Section 2.3. It is a combination of methods that are not quite as fast as Newton's method but that are very reliable. We can only use `fzero` to find roots of $f(x) = 0$ where f changes sign. There are two ways to use `fzero`. The first way is with a single starting value x_0 that you feel is fairly close to the desired root. This may not lead to convergence if the desired root is very near a singularity of the function. The call is `fzero(f, x0)`.

A second, and better, way to use `fzero` is to give it a "bracket" where f changes sign. Looking at the graph of f of Example 7.3, we see that to capture the root near $x = 2$, we can choose a bracket $[1, 3]$.

```
f = inline('sin(x) - .5*x')
root = fzero(f, [1 3]);
root = 1.89549426703398
```

`fzero` continues its iterations until it estimates that $|x_n - x_*| \le 2 \times 10^{-16}$. This is the default error tolerance. It is possible to require a different tolerance by adding another argument, depending on which version of MATLAB you are using. See the online help for `fzero`.

7.3   Solving a single equation in two variables

A single equation $f(x, y) = 0$ can sometimes be solved for y as a function of x with the symbolic solver. For example,

```
syms x y
f = x - exp(x*y);
solve(f,y)
ans =
log(x)/x
```

If we cannot solve the equation symbolically, we can make a numerical approximation. Suppose that we are given a point (x_0, y_0) such that $f(x_0, y_0) = 0$, and that $f_y(x_0, y_0) \ne 0$. Then by the implicit function theorem we know that for x near x_0 there exists a unique solution y of $f(x, y) = 0$. This means that for x near x_0, the level curve $f(x, y) = 0$ is the graph of a function $y = g(x)$. To find the values of y as a function of x numerically, we must numerically approximate a root of $y \to f(x, y) = 0$ for each x. We can do this with Newton's method or with a solver like `fzero` and a for loop.

Example 7.4

Let $f(x, y) = \sin(x) + y \exp(xy)$. We see that $f(0, 1) = 1$ and $f_y(0, 1) = 1$. To get an idea of what the level curve through $(0, 1)$ looks like, we can use the contour command. Put a mesh on the rectangle $[-1, 1] \times [0, 4]$. See Figure 7.4.

We see that we should be able to solve for y in terms of x for $x \geq -.2$. We could start at $(x, y) = (0, 1)$, but it may be easier to program to start at $x = -.2$. A look at the graph and some experimenting shows that the point $(x_1, y_1) = (-.2, 1.8)$ is almost on the curve, since $f(-.2, 1.8) = .0571$. f is defined in the mfile f.m. The script uses fzero (with a longer call) and numerically approximates the value of y on the curve for $-.2 \leq x \leq 1$, in steps of .01.

```
function z = f(y,x)
    z = sin(x) + y.*exp(x.*y)-1;
```
. . . .
```
% We choose the x values at which we will
% numerically estimate y
x = -.2:.01:1;

% We set aside the memory spaces for the y values.
y = zeros(size(x));

% Using our guess we calculate the first y value.
y(1) = fzero('f', 1.8, [], [], -.2);
```

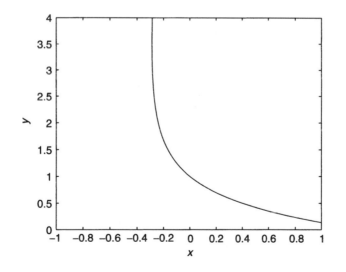

Figure 7.4 Level curve $f(x, y) = \sin(x) + y \exp(xy) - 1 = 0$.

```
% Now we enter a loop to calculate the
% remaining y values.
for n = 2:121
    y(n) = fzero('f', y(n-1), [], [], x(n));
end
plot(x,y)
```

Several comments are in order. The function function `fzero` operates on functions of one variable only. It can operate on a function of two or more variables if the other variables are regarded as parameters. Thus if $f = f(x, p_1, p_2)$ is given in an mfile as z = `f(x,p1,p2)`, then values of the parameters p_1, p_2 are passed to f with an augmented call to `fzero`, depending on which version of MATLAB you are using.

In versions prior to 5.3, the call is `fzero('f', y0, [], [], p1,p2)`. The blank brackets `[], []` in the call save places for additional information that can be given to `fzero`.

In version 5.3 `fzero` has been "improved." The call is instead `fzero('f', y0, [], p1, p2)`. Notice that we only need one place keeper.

In both cases, `fzero` regards the first argument of f as the "active" one and later arguments as the parameters. It is for this reason that in our function mfile `f.m` we wrote $z = f(y, x)$, reversing the order of the variables.

Finally note that we used the option of `fzero` that requires only a single starting value instead of a bracket. After finding the first value y_1, we entered the loop and used the previous value found for y as the first guess for the next value of y.

7.4   Newton's method in two dimensions

Newton's method in two dimensions is derived in the same manner as the version in one dimension. In one dimension, we start at a point x_1 and approximate the function f with the tangent line $l(x) \equiv f(x_1) + f'(x_1)(x - x_1)$. We then take the root x_2 of the equation $l(x) = 0$ as a better approximation to the root. The equation $l(x_2) = 0$ can be written

$$f'(x_1)x_2 = f'(x_1)x_1 - f(x_1). \tag{7.4}$$

We usually divide through the equation by $f'(x_1)$ to arrive at Eq. (7.3).

Now when we have a pair of equations, Eqs. (7.1), we make a similar linear approximation. We start with a first guess at the root, $p_1 = (x_1, y_1)$. We then approximate f by the plane tangent to the graph of f at $(p_1, f(p_1))$. As we saw in Eq. (5.8), this approximation is given by

$$l_f(x, y) \equiv f(p_1) + f_x(p_1)(x - x_1) + f_y(p_1)(y - y_1).$$

Similarly, the tangent plane approximation to the graph of g at $(p_1, g(p_1))$ is given by

$$l_g(x, y) \equiv g(p_1) + g_x(p_1)(x - x_1) + g_y(p_1)(y - y_1).$$

We replace system (7.1) by the linearized system

$$l_f(x, y) = 0 \tag{7.5}$$

$$l_g(x, y) = 0.$$

The terms in these equations can be rearranged a bit, in the same way as we arrived at Eq. (7.4):

$$f_x(p_1)x + f_y(p_1)y = f_x(p_1)x_1 + f_y(p_1)y_1 - f(p_1) \tag{7.6}$$

$$g_x(p_1)x + g_y(p_1)y = g_x(p_1)x_1 + g_y(p_1)y_1 - g(p_1).$$

Now this inhomogeneous linear system must be solved for a solution pair $p_2 = (x_2, y_2)$. The system may be thought of as the equations of two lines in the (x, y) plane, and we are seeking the point of intersection p_2 of these two lines.

System (7.6) can be put in a more compact form that will remind us even more of Eq. (7.4). We let $F(x, y)$ denote the pair of functions f, g:

$$F(x, y) = (f(x, y), g(x, y)).$$

Further, we let $J(p_1)$ denote the 2×2 matrix of partial derivatives

$$J(p_1) = \begin{bmatrix} f_x(p_1) & f_y(p_1) \\ g_x(p_1) & g_y(p_1) \end{bmatrix}.$$

J is called the *Jacobian matrix* of F at the point p_1. The first row of J is the gradient vector $\nabla f(p_1)$, and the second row is $\nabla g(p_1)$. Writing $p = (x, y)$ and $F(p) = (f(p), g(p))$ as column vectors we can express system (7.6) as

$$J(p_1)p = J(p_1)p_1 - F(p_1).$$

This is the analog of Eq. (7.4). The solution of this system is $p_2 = (x_2, y_2)$. Hopefully, it is a better approximation to the solution of the original system (7.1). The iteration scheme for Newton's method in two dimensions is therefore

$$J(p_n)p_{n+1} = J(p_n)p_n - F(p_n). \tag{7.7}$$

For computation, we break this down as follows:

1) Solve the linear system of equations $J(p_n)s = -F(p_n)$. s is called the Newton step.

2) Compute the next iterate, $p_{n+1} = p_n + s$.

Recall that it was very important to choose a good starting value for Newton's method in one dimension. This is even more true in two dimensions. One should always make a contour map of the zero curves of f and g to get an approximate location of the roots.

Example 7.5

We return to the system considered in Example 7.2, where $f(x, y) = y - \sin(x)$, and $g(x, y) = y - x^2 + 1$. For this pair of functions, the Jacobian matrix is

$$J(x, y) = \begin{bmatrix} f_x & f_y \\ g_x & g_y \end{bmatrix} = \begin{bmatrix} -\cos(x) & 1 \\ -2x & 1 \end{bmatrix}.$$

Here is a script that implements Newton's method for this pair of functions. The script is almost identical with the script for Newton's method discussed in Section 7.2. We must first write mfiles for $f(x, y)$, $g(x, y)$, and $J(x, y)$. The function mfile bigj.m produces a 2×2 matrix as output:

```
function z = f(x,y)
    z = y-sin(x);

function z = g(x,y)
    z = y - x.^2 +1;

function z = bigj(x,y)
    z = [-cos(x) 1; -2*x 1];
```

. . . .

```
p = input('enter the starting point p1 = [x1;y1]  ');
x = p(1);   y = p(2);

disp('  iterate      x      y      f(x,y)      g(x,y)  ')
[1, x, y, f(x,y), g(x,y)]

for n = 2:N
    s = -bigj(x,y)\[f(x,y); g(x,y)];
    p = p+s;
    x = p(1); y = p(2);
```

```
        [n,   x,    y,    f(x,y),   g(x,y)  ]
  end
```

Comments: The input for p has a semicolon between the entries. This makes p a 2×1 column vector. Similarly, in the first line of the loop we write [f(x,y); g(x,y)] with a semicolon to make this a column vector.

This program does not save the values of the iterates. The input statement puts the starting values $p_1 = (x_1, y_1)$ in the memory space p. The first time the loop is entered, the values p = [x1;y1] are used on the right side of the first line of the loop. The result of the calculation, which is the second iterate, $p_2 = (x_2, y_2)$, is placed in the same memory space p, overwriting the values that were there before. The results are printed out, and the loop continues. Each time, the latest iterate p_n is stored in the space p, overwriting the previous iterate, p_{n-1}.

The line s = -bigj(x,y)\[f(x,y);g(x y)] solves the system $J(p_n)s = -F(p_n)$ using the linear solve command x = A\b of Section 1.4. For reasons of accuracy and stability, it is usually better to solve a linear system with the A\b operation.

mfile newton2

The script mfile newton2.m is essentially the script written in the preceding subsection. The user must provide mfiles f.m, g.m, and bigj.m. The script plots the location of each iteration and pauses, waiting for the user to hit Return. This can be done for up to five iterations. Of course this script can be modified using a while loop and a required tolerance as seen in Chapter 13.

As in the one-dimensional case, it is important to know how close the approximate root is to the exact root. If $p_* = (x_*, y_*)$ is the exact root and $p_n = (x_n, y_n)$, we have the estimate

$$|x_n - x_*| + |y_n - y_*| \approx |x_n - x_{n-1}| + |y_n - y_{n-1}|.$$

Exercises

1. Let $f(x, y) = y - 3x(x - 1)(x + 1)$ and $g(x, y) = (x^2)/4 + y^2 - 1$.

a) Write mfiles f.m for f and g.m for g.

b) Put a meshgrid [X,Y] on the square $[-2, 2] \times [-2, 2]$. Plot the zero-level curves of f and g with the commands

```
contour(X,Y,f(X,Y),[0 0])
hold on
contour(X,Y,g(X,Y),[0 0], 'r').
```

Count the number of roots and estimate their location from the plot.

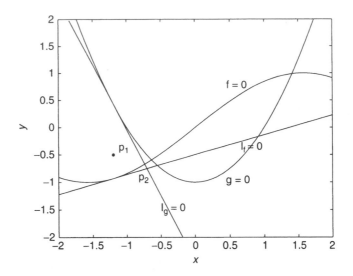

Figure 7.5 Zero level curves of $f(x, y) = y - \sin(x)$ and $g(x, y) = y - x^2 + 1$. Zero level lines of $l_f(x, y)$ and $l_g(x, y)$ for $p_1 = (-1.2, -.5)$. The intersection at p_2 is the next Newton iterate.

c) Define the symbolic expressions `ff = y -3*x*(x-1)*(x+1)` and `gg = .25*x^2 +y^2 -1`. Then use the symbolic solver as in Example 7.1. Does the symbolic solver get all the roots? Are they symbolic answers, or are they numeric answers?

2. Let $f(x, y) = y - 2\sin(2x)$. Use the same function g as in Exercise 1. Repeat parts a), b), and c) of Exercise 1.

3. Let $f(x, y) = \arctan(x+y) - xy$. The MATLAB function for arctan is `atan(x)`. Verify that $f(0, 0) = 0$.

a) Put a mesh on the square $[-1, 1] \times [-1, 1]$. Use the command `contour(X,Y,f(X,Y),[0 0])` to plot the zero level curve of f in this square. Is this curve the graph of a function $y = g(x)$? What is $f_y(0, 0)$?

b) Following Example 7.4, compute the values of $g(x)$ for $-1 \le x \le .5$, and plot the curve of values you get.

★   **4.** Consider the system

$$
\begin{aligned}
f(x, y) &= \arctan(x + y) - xy = 0 \\
g(x, y) &= y^2 - x = 0
\end{aligned}
$$

a) Write mfiles `f.m` and `g.m` for f and g.

b) There are three roots for this pair of equations. Plot the zero level curves of f and of g in the region $-2 \leq x, y \leq 2$ to get an estimate of the location of the roots.

c) Now write an mfile `bigj.m` as in Example 7.5 for the Jacobian matrix J. Then use the mfile `newton2.m` to get better results for each of the roots.

★ **5.** We wish to find the roots of the system

$$f(x, y) = x^2 + 4xy - 4y^2 - .1 = 0$$
$$g(x, y) = 4\cos x + 2\sin y - 5y = 0.$$

a) Plot the zero-level curves of f and g in the square $-3 \leq x, y \leq 3$. Make a first estimate of the location of the roots.

b) Use the mfile `newton2` to find the roots.

★★ **6.** A continuously stirred tank reactor is a vessel through which flows a mixture of chemicals. They react while in the vessel. We shall take the volume of the vessel to be V and the flow rate to be Q. We shall consider a reaction between three chemicals, A, B, and C, with the following reaction rules:

$$A \rightarrow 2B, \quad \text{with reaction rate } r_1,$$
$$B \rightarrow A + C, \quad \text{with reaction rate } r_2.$$
$$C \rightarrow A, \quad \text{with reaction rate } r_3.$$

Let A_0 be the inflow concentration of chemical A. Making an abuse of notation, let A be the concentration of chemical A in gram-moles/liter, and the same for B and C. The differential equations for the reaction process are

$$V\frac{dA}{dt} = Q(A_0 - A) - r_1 + r_2$$

$$V\frac{dB}{dt} = -QB + 2r_1 - r_2$$

$$V\frac{dC}{dt} = -QC + r_2 - r_3.$$

If there is a steady-state solution for this reaction, the concentrations of A, B, and C must satisfy

$$f_1(A, B, C) = Q(A_0 - A) - r_1 + r_2 = 0$$
$$f_2(A, B, C) = -QB + 2r_1 - r_2 = 0$$
$$f_3(A, B, C) = -QC + r_2 - r_3 = 0.$$

The reaction rates may depend on the concentrations in a nonlinear fashion. We shall assume

$$r_1 = .1A^2, \qquad r_2 = .04B^{3/2}, \qquad r_3 = .06C.$$

Thus a steady state reaction will correspond to a solution of the nonlinear system

$$
\begin{aligned}
f_1(A, B, C) &= Q(A_0 - A) - .1A^2 + .04B^{3/2} = 0 \\
f_2(A, B, C) &= -QB + .2A^2 - .04B^{3/2} = 0 \\
f_3(A, B, C) &= -QC + .04B^{3/2} - .06C = 0.
\end{aligned}
$$

Solve this system for $Q = 50$ liters/sec and $A_0 = 5$ gram-moles/liter.

a) Modify the file `newton2.m` to make it solve systems of three equations in three unknowns; call it `newton3.m`.

b) Make a first guess for the solution by assuming $B = C = 0$ and finding the value of A that solves the first equation. Then use `newton3.m` to solve the system.

★★   **7.** If a parameter is present in a system of equations such as system (7.1), the nature of the solutions can depend on the value of the parameter. Consider the system

$$
\begin{aligned}
f(x, y) &= (y - x)^2/4 - (x + y)/2 + 1 = 0 \\
g(x, y) &= y + \lambda(x - 2)^3 = 0.
\end{aligned}
$$

a) Use the command `contour` to display the level curves $f = 0$ and $g = 0$ for various values of λ in the rectangle $\{0 \le x \le 3, \ -.5 \le y \le 2\}$. Show that there is a critical value of λ, λ_*, such that

> there are no solutions for $\lambda < \lambda_*$;
> there is one solution for $\lambda = \lambda_*$;
> there are two solutions for $\lambda > \lambda_*$.

b) When $\lambda = 2$, the system has two roots. Use the code `newton2`, or use the symbolic solver, to find each of them.

c) Find the critical value λ_* and the single root of the system that corresponds to $\lambda = \lambda_*$. Use the following idea.

From the graphs of part a), we can see that when $\lambda = \lambda_*$, the level curves $C_1 = \{(x, y) : f(x, y) = 0\}$ and $C_2 = \{(x, y) : g(x, y, \lambda_*) = 0\}$ are tangent at the root (x_*, y_*). Let C_1 be the graph of $y_1(x)$ and C_2 be the graph of $y_2(x)$. By implicit differentiation,

$$-\frac{f_x(x_*, y_*)}{f_y(x_*, y_*)} = \frac{dy_1}{dx}(x_*)$$

$$= \frac{dy_2}{dx}(x_*) = -\frac{g_x(x_*, y_*, \lambda_*)}{g_y(x_*, y_*, \lambda_*)},$$

whence

$$f_x(x, y)g_y(x, y, \lambda) - f_y(x, y)g_x(x, y, \lambda)\Big|_{(x_*, y_*, \lambda_*)} = 0.$$

Thus (x_*, y_*, λ_*) is the solution of the system of three equations

$$f = 0$$

$$g = 0$$

$$f_x g_y - f_y g_x = 0 .$$

Solve this system using the three-dimensional version of the Newton code or the symbolic solver.

8. Suppose a weight of mass m is suspended by two springs, attached at points $(-1, 0)$ and $(1, 0)$. What are the coordinates (x, y) of the position of the mass? This situation is shown in Figure 7.6.

We assume a linear spring law, in which the magnitude of the restoring force is proportional to the amount the spring is stretched. To be more specific, we suppose that the spring on the left has length l_1 when not stretched and has spring constant k_1, while that on the right has length l_2 when not stretched and spring constant k_2. With these assumptions, the magnitude of the force exerted by the spring on the left when it is stretched to a length $d_1 > l_1$ is

$$\|\mathbf{F}_1\| = k_1(d_1 - l_1).$$

Similarly,

$$\|\mathbf{F}_2\| = k_2(d_2 - l_2).$$

The forces $\mathbf{F}_1, \mathbf{F}_2$ are

$$\mathbf{F}_1 = \|\mathbf{F}_1\|(-\cos(\theta_1), \sin(\theta_1))$$

and

$$\mathbf{F}_2 = \|\mathbf{F}_2\|(\cos(\theta_2), \sin(\theta_2)),$$

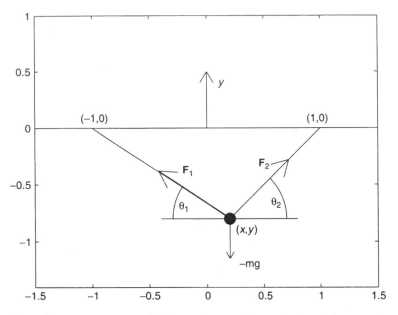

Figure 7.6 Weight with mass m suspended by springs with gravitational force and restoring forces of springs.

with

$$\cos(\theta_1) = (x+1)/d_1, \qquad \sin(\theta_1) = -y/d_1, \qquad d_1 = \sqrt{(x+1)^2 + y^2}$$

and

$$\cos(\theta_2) = (1-x)/d_2, \qquad \sin(\theta_2) = -y/d_2, \qquad d_2 = \sqrt{(x-1)^2 + y^2}.$$

At equilibrium, the sum of forces acting on the mass is zero:

$$\mathbf{F}_1 + \mathbf{F}_2 - mg\mathbf{j} = 0.$$

Writing out the x and y components separately, we have

$$0 = \mathbf{F}_{1,x} + \mathbf{F}_{2,x} = -k_1(1+x)(1-l_1/d_1) + k_2(1-x)(1-l_2/d_2)$$
$$mg = \mathbf{F}_{1,y} + \mathbf{F}_{2,y} = -k_1 y(1-l_1/d_1) - k_2 y(1-l_2/d_2).$$

a) Show that if $l_1 = l_2 = l$ and $k_1 = k_2 = k$, then $x = 0$, so $d_1 = d_2 = \sqrt{1 + y^2}$. In this case, the system reduces to the second equation, now an equation in just y. Set $k = l = 1$ and $g = 9.8$. Write an mfile that takes the mass m as input and solves numerically for y. Keep in mind that with our coordinate system, $y < 0$.

b) Now rewrite the mfile of part a) so that it calculates the root y for each value of $m = 1, 2, 3, \ldots, 50$ and plots the values of y against the values of m. This graph can be well approximated by a linear function $y = c_1 m + c_2$. What choices of c_1 and c_2 give the best linear approximation? Explain the best choice of c_1 by examining the equation for y.

c) For simplicity set $k_1 = 1, l_1 = 1, m = 1, g = 9.8$. Write an mfile that solves the system with input statements for the values of k_2 and l_2 and uses the numerical solver newton2. The mfile should also plot out the position of the mass as shown in Figure 7.6.

8

Optimization

Prepared mfiles used in this chapter

```
findcrit laqrange impl
```

8.1 Critical points and the second-derivative test

Recall that a critical point for a function $f(x, y)$ is a point (x_0, y_0) in its domain where f is differentiable, with $f_x(x_0, y_0) = 0$ and $f_y(x_0, y_0) = 0$. Geometrically this means that the tangent plane to the graph of f at the point $(x_0, y_0, f(x_0, y_0))$ is parallel to the x, y plane ("horizontal"). A critical point for f can be a local maximum, a local minimum, or neither of these. Figure 8.1 displays the graph of a function having two critical points with attached tangent planes.

To find the critical points of f we must solve the system

$$f_x(x, y) = 0 \qquad (8.1)$$

$$f_y(x, y) = 0.$$

Solutions of these equations (there may be many) can sometimes be found using a symbolic solver and sometimes using a numerical solver. We shall deal with this issue later in the chapter.

Since the first derivatives of f are zero at (x_0, y_0), we must look to the higher derivatives to describe the behavior of f near a critical point. The contour lines can also provide valuable information.

Recall that the second-derivative test for functions of one variable says that if $f'(x_0) = 0$ and $f''(x_0) > 0$, then x_0 is a local minimum for f. If $f''(x_0) < 0$, then x_0 is a local maximum for f.

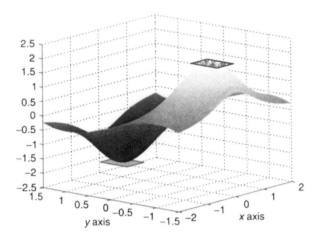

Figure 8.1 Graph of $f(x, y) = (y^3 - 3y)/(1 + x^2)$, with horizontal tangent planes at the critical points $(0, -1)$ and $(0, 1)$.

How can we extend this idea to higher dimensions? Intuitively, if $f(x, y)$ has a local minimum at the point (x_0, y_0), then the intersection of the graph of f with any vertical plane,

$$a(x - x_0) + b(y - y_0) = 0,$$

should be a curve that is concave up near the point (x_0, y_0). See Figure 8.2. If (x_0, y_0) is a maximum, we would expect the curve of intersection to be concave down near (x_0, y_0).

The second-derivative test for functions of two variables gives a sufficient condition for this kind of behavior. It will also give us information about the level curves of a function near a local maximum or minimum. We define the *discriminant*

$$D(x, y) = f_{xx}(x, y) f_{yy}(x, y) - f_{xy}^2(x, y). \tag{8.2}$$

D is the determinant of the 2×2 matrix of second-order partial derivatives

$$\mathbf{H}(x, y) = \begin{bmatrix} f_{xx}(x, y) & f_{xy}(x, y) \\ f_{xy}(x, y) & f_{yy}(x, y) \end{bmatrix}. \tag{8.3}$$

H is known as the *Hessian matrix* of f and can be thought of as the analog of the second derivative of a function of one variable.

The Second-Derivative Test

- If $D(x_0, y_0) > 0$ and $f_{xx}(x_0, y_0) > 0$, then (x_0, y_0) is a local minimum.
- If $D(x_0, y_0) > 0$ and $f_{xx}(x_0, y_0) < 0$, then (x_0, y_0) is a local maximum.

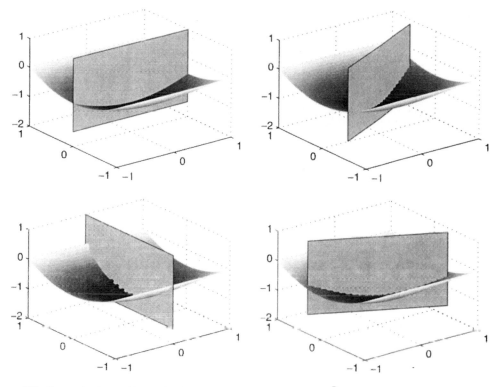

Figure 8.2 Several slices through the graph of $f(x, y) = (y^2 - 1) \cos(x)$ at the critical point $(x_0, y_0) = (0, 0)$.

- If $D(x_0, y_0) < 0$, then (x_0, y_0) is a *saddle point*.
- If $D(x_0, y_0) = 0$, then we cannot draw a conclusion.

To understand how the second-derivative test works, we apply it to a quadratic function,

$$f(x, y) = ax^2 + 2bxy + cy^2, \tag{8.4}$$

that has a critical point at $(x_0, y_0) = (0, 0)$. For this f, the quantity $D(x, y) \equiv 4(ac - b^2)$. Now suppose $D > 0$ and $2a = f_{xx}(x_0, y_0) > 0$. Then from analytic geometry, we know that the graph of f is a bowl opening upward with minimum at $(0, 0)$. When $D > 0$ and $a < 0$, the graph is a bowl opening downward. In both cases the level curves of f are ellipses centered at $(0, 0)$.

We illustrate the case $D < 0$ with an example. Set $a = 1, b = 2$, and $c = -2$ so that $D = -24$. The graph of f is displayed in Figure 8.3 sliced by planes parallel

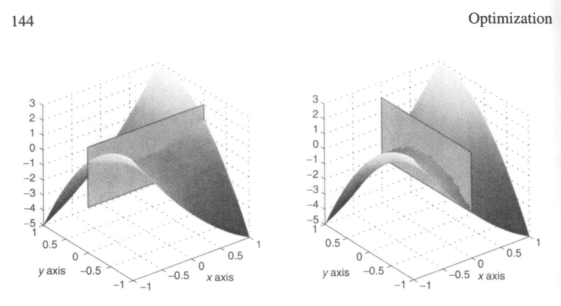

Figure 8.3 Graph of $f(x, y) = x^2 + 4xy - 2y^2$ near saddle point $(x_0, y_0) = (0, 0)$.

to the x, z plane and to the y, z plane. The level curves of f are hyperbolas, shown in Figure 8.4.

For a function f more general than Eq. (8.4) we approximate f at a critical point by a quadratic function. We use the two-dimensional version of the Taylor

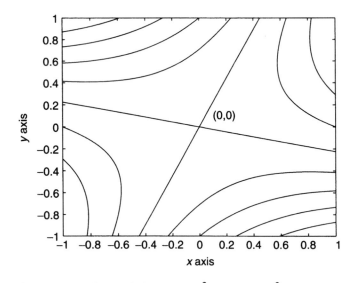

Figure 8.4 Contour lines of $f(x, y) = x^2 + 4xy - 2y^2$ near saddle point.

expansion at a point (x_0, y_0):

$$f(x, y) = f(x_0, y_0) + f_x(x_0)(x - x_0) + f_y(x_0, y_0)(y - y_0) + q(x, y) + E, \quad (8.5)$$

where $q(x, y)$ is the quadratic function

$$q(x, y) = +\frac{1}{2}[f_{xx}(x_0, y_0)(x - x_0)^2 + f_{yy}(x_0, y_0)(y - y_0)^2] \quad (8.6)$$

$$+2f_{xy}(x_0, y_0)(x - x_0)(y - y_0).$$

When f has continuous third-order partial derivatives, the error E can be estimated as

$$|E| \leq C \max |x - x_0|^3, |y - y_0|^3, \quad (8.7)$$

where the constant C is determined by the behavior of the third derivatives near (x_0, y_0). We recognize the first three terms of Eq. (8.5) as the tangent plane approximation to f at (x_0, y_0). At a critical point, $f_x(x_0, y_0) = f_y(x_0, y_0) = 0$, so the Taylor expansion (8.5) reduces to

$$f(x, y) = f(x_0, y_0) + q(x, y) + E. \quad (8.8)$$

Since the error $E \to 0$ as $(x, y) \to (x_0, y_0)$, f behaves essentially like q near a critical point. This implies that for (x, y) close to (x_0, y_0), the level curves of f should be approximated by the level curves of q. Now, q is of the form (8.4), with

$$2a = f_{xx}(x_0, y_0), \qquad 2b = f_{xy}(x_0, y_0), \qquad 2c = f_{yy}(x_0, y_0)$$

and with x replaced by $x - x_0$ and y replaced by $y - y_0$. In the first two cases of the second-derivative test, when $D(x_0, y_0) > 0$, the level curves of q are ellipses, centered at (x_0, y_0). Examples are shown in Figures 8.5 and 8.6.

The third case, which has no one-dimensional analog, is that the quantity $D < 0$. This can happen when the two partial derivatives f_{xx} and f_{yy} are both nonzero and have differing signs or if both $f_{xx} = f_{yy} = 0$ at (x_0, y_0) and $f_{xy} \neq 0$ at (x_0, y_0). The graph of f near the critical point is called a *saddle surface* in this case. See Figure 8.3.

Of course, one can have critical points where $D = 0$ so that the second-derivative test does not apply. The graph of f and its level curves can be quite complicated near such a critical point, as shown in the following example.

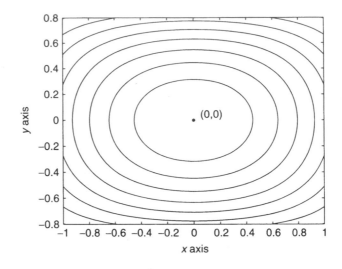

Figure 8.5 Level curves of $f(x, y) = (y^2 - 1)\cos(x)$ near a minimum at $(x_0, y_0) = (0, 0)$.

Example 8.1

Let $f(x, y) = x^3 - 3xy^2$. The origin is a critical point for f and $D(0, 0) = 0$. We display its graph and the level curves near $(0, 0)$ in Figure 8.7. We used a finer mesh for the contours to avoid irregular jumps in the level curve near the orgin.

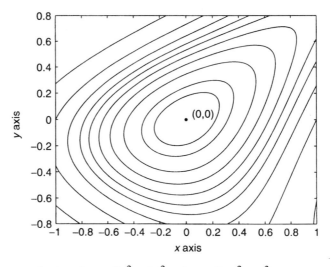

Figure 8.6 Level curves of $g(x, y) = -2x^2 - 3y^2 + 2xy + 3xy^2 - y^3$ near a maximum at $(x_0, y_0) = (0, 0)$.

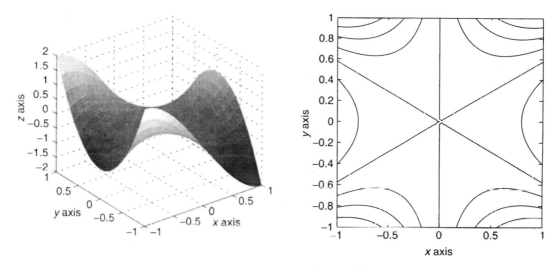

Figure 8.7 Graph of $f(x, y) = x^3 - 3xy^2$ near $(0, 0)$.

```
f = inline('x.^3 - 3*x.*y.^2', 'x', 'y')
% We plot the graph of f
  in the left side of the figure.
[X,Y] = meshgrid(-1:.1:1);
subplot(1,2,1)
    surf(X,Y,f(X,Y))
% We plot the contours
  in the right side of the figure.
[XX,YY] = meshgrid(-1:.025:1);
subplot(1,2,2)
    contour(XX,YY,f(XX,YY), 'k')
```

8.2 Estimating the maximum and minimum

MATLAB has a very convenient tool that can be used to calculate the maximum
and minimum over a mesh. If $\mathbf{v} = [v_1, \ldots, v_n]$ is a vector, either row or column,
max(v) returns max v_i. min(v) returns min v_i. The more complete command [V,i]
= max(v) returns max v_i and the first index i where the maximum is attained.

Example 8.2

```
>> v = [1 2 -1 3 0 3];
>> V = max(v);
```

```
V =
        3
>> [V,i] = max(v);
V =
        3
i =
        4
```

We can use this tool to find the maximum or minimum value of a function of one variable over a mesh. Here we estimate the maximum of $f(x) = x^2(1 - x^2)$ over the interval $[-1, 1]$:

```
>> f = inline('x.^2.*(1-x.^2)')
>> x = -1:.02:1;
>> [M,i] = max(f(x))
M =
        .2499
i =
        16
>> x0 = x(i)
x0 =
        -.7000
```

It is important to keep in mind that this sequence of commands has found the maximum of f over the mesh, not over the interval $[-1, 1]$. The true maximum of f over $[-1, 1]$ is .25, and it is attained at two places, $x = \pm 1/\sqrt{2} = \pm.707161$.

To treat functions of two variables, we use the commands on matrices. If A is a matrix with elements $a_{i,j}$, the command max(A) finds the maximum in each column, recording the results in a row vector. Then repeating the command, max(max(A)), we find the maximum of the row vector max(A). Thus max(max(A)) is the maximum over the elements $a_{i,j}$. In the same way, min(min(A)) is the minimum over the elements $a_{i,j}$. The command [row, I] = max(A) returns the largest element in each column and the first index i in each column where the maximum is attained. Then applying max again, we can find the maximum over all the $a_{i,j}$ and a pair of indices where the maximum is attained.

Example 8.3

```
>> A = [1 0 3; 2 1 3; -1 4 0; 0 2 1]
A =
        1    0    3
        2    1    3
```

```
          -1   4   0
           0   2   1
          [row, I]  = max(A)
          row =
             2   4   3
          I =
             2   3   1

          [maxA, j]  = max(row)
          maxA =
                 4
          j =
                 2
          >> i = I(j)
          i =
                 3
```

Thus we see that the largest element of A is $A(3, 2) = 4$.

If f is a function of two variables, we can use the max command to estimate the maximum of f over a set G. Let $f(x, y) = x/2 + \exp(-x^2 - y^2)$ and $G = \{0 \le x \le 1\} \times \{-1/2 \le y \le 1/2\}$.

```
          >> f = inline('.5*x+exp(-x.^2-y.^2)','x','y')
          >> x = 0:.02:1;
          >> y = .5:.02:.5;
          >> [X,Y]  = meshgrid(x,y);
          >> [row, I]  = max(f(X,Y));
          >> [maxf, j]  = max(row)
          maxf =
                 1.0646
          j =
                14
          >> x0 = x(j)
          x0 =
                 .2600
          >> y0 = y(I(j))
          y0 =
                 0
```

Then $f(x_j, y_i) \le f(.26, 0) = 1.0646$ for all the points in the mesh.

The commands min and max can also be used on three-dimensional arrays. See help min or help max.

mfile findcrit

The mfile `findcrit` uses contour maps and the commands `max` and `min` to find estimates of the maximum and minimum of a function over a rectangle. The call is `findcrit(f, corners)` when the function f is given as an inline function and `findcrit('f',corners)` when f is given in an mfile. `corners` is a vector, [*a b c d*]. The corners of the rectangle are (a, c), (b, c), (b, d), and (a, d). After the call is made, a contour map of the rectangle is displayed. The script waits for the user to click twice with the mouse on the lower left and the upper right corners of a smaller rectangle where you think the critical point lies. A smaller blue rectangle will be displayed. Then click a third time, anywhere. The blue rectangle will be blown up to fill the figure, and it will contain a blown-up contour map. The maximum and minimum of f over the blue rectangle will be computed and displayed on the screen, along with the coordinates in the mesh where they are attained. This process can be repeated four times.

We now give an example where several of these techniques are combined.

Example 8.4

Let
$$f(x, y) = \sin(x + y) + 1 - x^2/40 + e^{-y^2}.$$
We wish to find the critical points of this function in the rectangle $R = \{-6 \le x \le 9, \ -4 \le y \le 4\}$. First we graph f on R:

```
f= inline('sin(x+y) +
        1 - (1/40)*x.^2 + exp(-y.^2)', 'x', 'y')
[X,Y] = meshgrid(-6:.2:9,-4:.2:4);
surf(X,Y,f(X,Y))
```

The graph is shown in Figure 8.8. It appears that there are a number of hills and valleys in R. Next we use a contour map on R to get a rough location of the critical points in R:

```
contour(X,Y,f(X,Y),20)
```

In Figure 8.9, we see that local maxima alternate with saddle points. To find the coordinates of these critical points, we must solve Eq. (8.1) or use graphical techniques. First we shall try the symbolic equation solver on equations (8.1).

```
syms x y
fx = cos(x+y) - x/20;
```

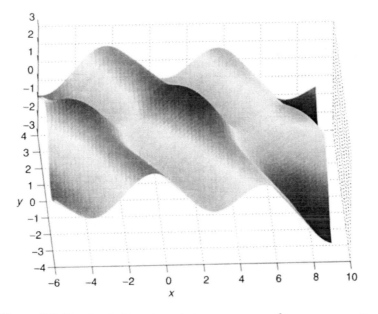

Figure 8.8 Graph of $f(x, y) = \sin(x + y) + 1 - x^2/40 + \exp(-y^2)$.

```
fy = cos(x+y) - 2*y*exp(-y^2)
[a,b] = solve(fx,fy)
double([a,b]);
ans =
1.46109980190872   0.03657639542810
```

We see that the symbolic solver has found only one critical point, and has found it numerically. Now to apply the second-derivative test, we calculate the discriminant symbolically and evaluate it at the critical point:

```
fxx = diff(fx,x); fyy = diff(fy,y)
fxy = diff(fx,y);
D = fxx*fyy - fxy^2;
subs(D,[x,y],[a,b])
ans =
2.13612469355530508460542646650197

subs(fxx, [x,y], [a,b])
ans =
-1.04732791418978931786442401750e64
```

Since $D > 0$ and $f_{xx} < 0$ at the critical point, we conclude that it is a local maximum, in agreement with our graphical results in Figures 8.8 and 8.9.

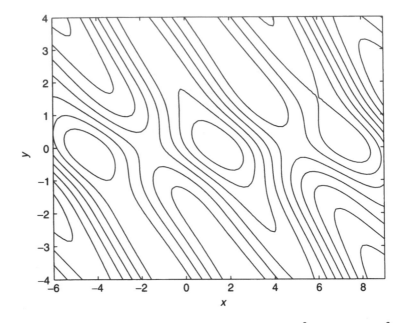

Figure 8.9 Contours of $f(x, y) = \sin(x + y) + 1 - x^2/40 + \exp(-y^2)$.

To find the other critical points in R, we shall use the mfile `findcrit`. From Figure 8.9, there appears to be a saddle point near $(-1, 0)$. To close in on this saddle point, we use `findcrit` with `corners` chosen to be $[-2, 0, -1, 1]$. Then after clicking several times we have the picture shown in Figure 8.10. The saddle point appears to be located at $(a, b) = (-1.61, -.040)$. To confirm this graphical approach, we evaluate f_x, f_y and D at this point:

```
a = -1.61; b = -.040
double(subs(fx, [x,y],[a,b]))
ans =
  -1.1158e-04
double(subs(fy, [x,y], [a,b]))
ans =
  -2.9775e-04
double(subs(D, [x,y], [a,b]))
ans =
  -1.9341
```

Of course, these results are only quite approximate. We could repeat the procedure with `findcrit` and a much smaller starting rectangle to get a more accurate location of the critical point. We could also use Newton's method to find a solution

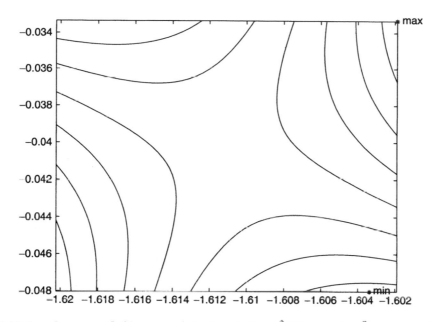

Figure 8.10 Level curves of $f(x, y) = \sin(x + y) + 1 - x^2/40 + \exp(-y^2)$ near a saddle point.

of system (8.1). We would need mfiles for f_x, f_y, and the Jacobian matrix of system (8.1), which would be the Hessian matrix $\mathbf{H}$ of f. For a starting point, we could use the approximate location of the critical point found using findcrit.

8.3 Constrained maximum and minimum problems

It is an important theorem of analysis that a continuous function on a closed, bounded set in R^2 or R^3 attains its maximum and minimum. We can find the location of an extreme point by seeking the critical points if that extreme point occurs in the interior of the set. If the maximum or the minimum occurs on the boundary, it may not be the case that the first derivatives vanish there.

We suppose the set is described as

$$K = \{(x, y) : g(x, y) \le 0\}$$

and the boundary of the set K is the level curve $\{(x, y) : g(x, y) = 0\}$. The constrained max-min problem is to find the maximum or minimum of a function $f(x, y)$ over the set K. This is often stated as follows:

Find max f or min f subject to the constraint $g(x, y) \le 0$.

The maximum or minimum may occur in the interior of K or on the boundary of K.

The method of Lagrange multipliers is used to find the extrema of functions on the boundary curve of a set. To find the minimum and maximum values of $f(x, y)$ on the boundary of K we look for those points on the boundary where a level curve of f is tangent to the boundary of K. Since the gradient vector is always orthogonal to the level curves, we see that at such a point, the gradients of f and g point in the same direction or in the opposite direction. See Figure 8.11.

A concise way to state this is to say that at these points,

$$\nabla f(x_0, y_0) = \lambda \nabla g(x_0, y_0)$$

for some scalar λ. λ is a third variable, called the *Lagrange multiplier*. When we write out the components of the vector equation and add the constraint that (x_0, y_0) must lie on the boundary curve $g(x, y) = 0$, we arrive at a system of three equations in the three variables x, y, and λ:

$$f_x(x, y) = \lambda g_x(x, y) \tag{8.9}$$

$$f_y(x, y) = \lambda g_y(x, y) \tag{8.10}$$

$$g(x, y) = 0. \tag{8.11}$$

The solutions of this system are candidates for the location of the maximum and the minimum of f on the curve $g(x, y) = 0$. This system, like system (8.1), can

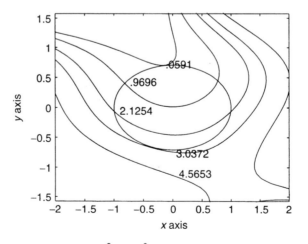

Figure 8.11 Boundary curve $g(x, y) = x^2 + 2y^2 - 1 = 0$ and level curves of $f(x, y) = x^2 + (y - 1)^2 + xy^3$.

be solved symbolically sometimes and numerically sometimes. Solutions of these systems are discussed in the next section.

The following mfile `lagrange` can be used to get a better geometric picture of the level curves and can be used to get an estimate of the solutions of Eqs. (8.9)–(8.11). For example, in Figure 8.11, which was produced by the mfile `lagrange.m`, the maximum of f on the curve $g = 0$ is approximately 3.04, while the minimum value is approximately .05. Since there are no interior critical points, these are the absolute maximum and minimum of f over the set $K = \{g(x, y) \le 0\}$.

mfile lagrange

The mfile `lagrange.m` has the call `lagrange(f,g,corners)` when the functions f and g are given as inline functions and `lagrange('f', 'g',corners)` when given in mfiles `f.m` and `g.m`. As in the mfile `findcrit`, the vector corners is $[a\ b\ c\ d]$, where (a, c), (b, c), (b, d), and (a, d) are the corners of the viewing rectangle. After the call, the level curve $g(x, y) = 0$ is displayed. The script waits for the user to click on a point anywhere in the viewing rectangle. When the mouse is clicked, the level curve of f through that point is plotted, as well as the value of f. This can be done five times. By marching around the curve, we can come close to the point where the maximum or minimum on the boundary curve is attained.

Example 8.5

Let $f(x, y) = x^3 + x^2 + y^2/3$ and let K be the disk of radius 6 centered at the origin. $K = \{x^2 + y^2 \le 36\}$, so we take $g(x, y) = x^2 + y^2 - 36$. The system (8.9)–(8.11) to solve is

$$3x^2 + 2x = 2\lambda x$$
$$2y/3 = 2\lambda y$$
$$x^2 + y^2 = 36.$$

First we graph several of the level curves of f using the mfile `lagrange.m`. Then we solve the system symbolically:

```
f = inline('x.^3 + x.^2 + (y.^2)/3', 'x', 'y')
g = inline('x.^2 + y.^2 -36', 'x', 'y')
corners = [- 8,9.5,-7,7]
lagrange(f,g,corners)
several clicks
```

```
% now we begin the symbolic calculation of the roots.
syms x y lambda
eq1 = 3*x^2 + 2*x - 2*lambda*x
eq2 = (2/3)*y - 2*lambda*y
g = x^2 +y^2 -36
[lambda x y ] = solve(eq1,eq2,g);
[lambda x y]
ans =
[                10,            6,                    0]
[                -8,           -6,                    0]
[               1/3,            0,                    6]
[               1/3,            0,                   -6]
[               1/3,         -4/9,      10/9*29^(1/2)]
[               1/3,         -4/9,     -10/9*29^(1/2)]
```

In agreement with the graph of the level curves in Figure 8.12, we see that the maximum is attained at $(x, y) = (6, 0)$ and the minimum at $(x, y) = (-6, 0)$. The remaining four points are local extrema of f on the curve $g = 0$. We can see from Figure 8.12 that the level curves of f are tangent to the curve $g = 0$ at those points as well.

It is often the case that the 3×3 system (8.9)–(8.11) cannot be solved symbolically. We may have to use numerical methods to find solutions. Newton's method can easily be extended to these larger systems. However, now it will be harder to find good starting values.

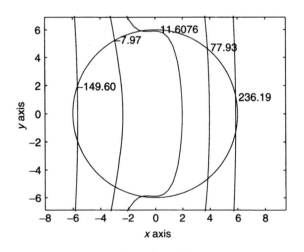

Figure 8.12 Level curves of $f(x, y) = x^3 + x^2 + y^2/3$ on and near the disk $\{x^2 + y^2 \leq 36\}$.

8.4   Functions of three variables

A critical point for a function $f(x, y, z)$ is a point where all the first partial derivatives vanish. This condition yields a system of three equations in three unknowns,

$$f_x(x, y, z) = 0 \qquad (8.12)$$
$$f_y(x, y, z) = 0$$
$$f_z(x, y, z) = 0.$$

The solutions of this system are candidates for a local maximum or minimum of f. This system may be solvable by symbolic means, but more often a numerical method must be used, which requires a reasonably good first guess for a starting point. Locating the critical points, even approximately, is not easy when it is not possible to graph the function. One must use all the information available from the setting of the problem to gain some insight. The commands slice and impl, discussed in Chapter 6, can be useful here.

Example 8.6

Let

$$f(x, y, z) = 1.5e^{-(x-1)^2-y^2-z^2} + e^{-(x+1)^2-(y-1)^2-z^2}.$$

From the form of the function we expect to find two local maxima, one near $(x, y, z) = (1, 0, 0)$ and one near $(x, y, z) = (-1, 1, 0)$. First we write an mfile for f:

```
function w = f(x,y,z)
    w1 = 1.5*exp(-(x-1).^2 - y.^2 - z.^2);
    w2 = exp(-(x+1).^2 - (y-1).^2 - z.^2);
    w = w1 + w2;
```

Then we construct a three-dimensional array and use the commmand slice as in Section 6.2:

```
x = linspace(-1.5, 1.5, 41);
y = linspace(-.5, 1.5, 41);
z = linspace(-1,1,41);
[X,Y,Z] = meshgrid(x,y,z);
W = f(X,Y,Z);
slice(X,Y,Z,W, [-1,0,1], 0,0); colorbar
```

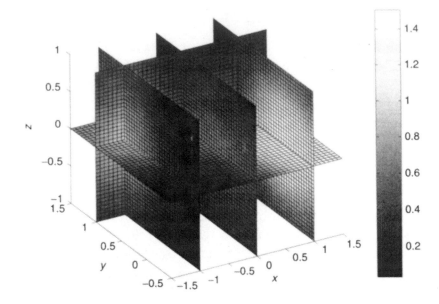

Figure 8.13 Slices through the domain of $f(x, y, z)$ indicating possible local maxima.

Note that we have substituted a vector $[-1, 0, 1]$ for the coordinate of the planes parallel to the x axis. This gives us slices through three parallel planes $x = -1$, $x = 0$, and $x = 1$. The result is shown in Figure 8.13.

The colorbar indicates that the largest values of f correspond to the lightest shade of gray (on this page). It appears that there are local maxima near $(1, 0, 0)$ and $(-1, 1, 0)$. To get good approximate values of these critical points, we would use Newton's method on Eq. (8.12).

Once a critical point has been found, what is the second-derivative test? This time the Hessian matrix is the 3×3 symmetric matrix of second-order partial derivatives,

$$\mathbf{H}(x, y, z) = \begin{bmatrix} f_{xx}(x, y, z) & f_{xy}(x, y, z) & f_{xz}(x, y, z) \\ f_{xy}(x, y, z) & f_{yy}(x, y, z) & f_{yz}(x, y, z) \\ f_{xz}(x, y, z) & f_{yz}(x, y, z) & f_{zz}(x, y, z) \end{bmatrix}.$$

A sufficient condition that (x_0, y_0, z_0) be a local minimum for f is that

1. $f_{xx} > 0$,
2. $f_{xx} f_{yy} - f_{xy}^2 > 0$,
3. $\det \mathbf{H} > 0$,

all evaluated at (x_0, y_0, z_0). Notice how this condition extends the first case of the second-derivative test for functions of two variables. If we apply this set of conditions to the function $-f$, we see that a sufficient condition for the critical point (x_0, y_0, z_0) to be a local maximum for f is that quantity (1) be negative, quantity (2) be positive, and quantity (3) be negative. In both of these cases, the level surfaces of f close to the critical point will resemble ellipsoids. We can see this with the command impl.

We continue to study the function of Example 8.6. From the form of the function we expect the local maximum near $(1, 0, 0)$ to be about 1.5, while the local maximum near $(-1, 1, 0)$ should be about 1. Thus the level surfaces S_c for $1 < c < 1.5$ should consist of a single closed surface resembling a sphere containing the point $(1, 0, 0)$, while for $c < 1$, but not too small, the level set S_c should consist of two closed surfaces, one containing $(1, 0, 0)$ and the other containing $(-1, 1, 0)$. Eventually for $c > 0$ sufficiently small, S_c will consist of one surface enclosing both critical points:

```
f = inline('1.5*exp(-(x-1).^2 - y.^2 - z.^2) +
  exp(-(x+1).^2 - (y-1).^2 - z.^2)', 'x', 'y', 'z')

corners = [-2,2,-2,2, -2,2];
subplot(2,2,1)
    impl(f, corners,  5)
    title('c = .5')
subplot(2,2,2)
    impl(f, corners, .9)
    title('c = .9')
subplot(2,2,3)
    impl(f, corners, .7)
    title('c = .7')
subplot(2,2,4)
    impl(f, corners, 1.2)
    title('c = 1.2')
```

See Figure 8.14.

Exercises

1. A function mfile need not produce numbers as output. It may produce a graph.

Write a short function mfile, conic.m, that takes as its three arguments a, b, c and then graphs the quadratic function $f(x, y) = ax^2 + 2bxy + cy^2$ over the square $-2 \le x, y \le 2$. It will begin like this:

```
function out = conic(a,b,c)
    [X,Y] = meshgrid(-2:.05:2);
    Z = . . . .
```

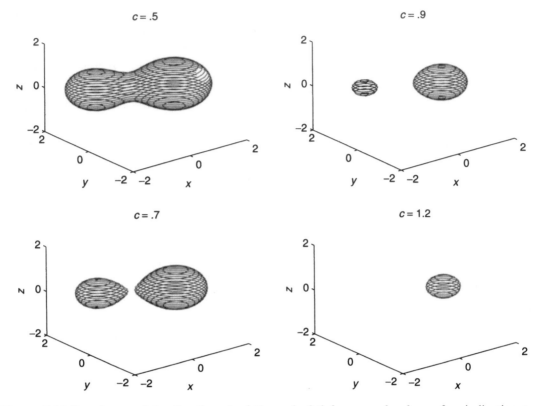

Figure 8.14 Level sets of the function f of Example 8.6 for several values of c, indicating two local maxima.

$$\texttt{surf(X,Y,Z); shading flat; colormap(cool)}$$

Add a statement that superimposes the plane $z = 0$ so that you can see clearly where f is positive and where negative.

Try the following choices of a, b, c. In each case describe the surface (bowl up, bowl down, or saddle) and the nature of the critical point.

(i) $(a, b, c) = (1, 0, 2)$ (ii) $(a, b, c) = (-2, 0, -3)$
(iii) $(a, b, c) = (1, 0, -2)$ (iv) $(a, b, c) = (0, 2, 0)$
(v) $(a, b, c) = (1, 3, 2)$ (vi) $(a, b, c) = (3, 2, 2)$

2. Let $g(x, y)$ be

$$g(x, y) = 2e^{-(x-1)^2-(y-1)^2} + 1.8e^{-5(x+1)^2-3y^2} - 2e^{-2(x-1)^2-3(y+.5)^2}.$$

a) Graph the function on the square $[-2, 2] \times [-2, 2]$. You will see that the graph of g has two hills and one valley.

b) Use the contour command `contour(X,Y,g(X,Y),20)`. This will plot 20 contour lines. Use the contour map to estimate the coordinates of the two peaks and of the bottom of the valley.

c) Use the mfile `findcrit` to explore a possible critical point near the point $(-.35, .45)$. What kind of critical point is this? Put a fine mesh over a small rectangle containing $(-.35, .45)$ and graph the function with the command `surf`.

3. Let $f(x, y) = \sin(x)/(1 + (y - x)^2)$.

a) Graph f over the square $[-3, 3] \times [-3, 3]$. You should see one hill and one valley.

b) Use the mfile `findcrit` to estimate where the maximum occurs. Then find the critical point by hand or by using the symbolic solver.

c) Use `findcrit` to determine where the minimum occurs.

4. Let $f(x, y) = x^2 + 2y^2 + 3xy^3 - y^3$.

a) Use the contour command to locate approximately the critical points of f.

b) Then solve system (8.1) symbolically for the critical points as follows: Define the symbolic expressions for f_x and f_y. Then use the command `[a b] = solve(fx fy)`, as in Example 8.4. The answer will fill the screen. To get a double-precision evaluation of these roots, use `double([a b])`.

5. Let $f(x, y) = x^4 - 2x^2 - y^3 + 3y$. This function has six critical points in the square $-2 \le x, y \le 2$.

a) Make a fairly fine mesh over the square, say, 50 by 50. Then use the commmand `contour(X,Y,Z, levels)`, where `levels` = `-4:.4:4`. Locate the six critical points and determine their nature by looking at the contour map.

b) Now solve system (8.1) for the critical points by hand. Use the second-derivative test at each critical point. Compare these results with your observations in part a).

★ **6.** Let

$$f(x, y) = x + y + 2e^{-(x-1)^2 - y^2} - 3e^{-(x+1)^2 - y^2}.$$

a) Make an mfile for f. Graph f over the square $\{-3 \le x, y \le 3\}$ using the command `surf`.

b) Make a contour map over the same square, using 50 contours. Locate where you think there are critical points.

c) Use the mfile `findcrit` to get good approximations to the location of the critical points, and determine their nature.

d) Write mfiles for f_x, f_y and **H**. Use the mfile `newton2` to get higher-accuracy approximations of the critical points.

7. An exercise on Taylor polynomials in one dimension.

a) Let $f(x) = e^x$. Let $p_0(x) \equiv 1$, $p_1(x) = 1+x$, and $p_2(x) = 1+x+x^2/2$ be the Taylor polynomials of f at $x = 0$ of degrees zero, one, and two, respectively. Plot f, p_0, p_1, and p_2 together on the interval $[-2, 2]$. Compute $\max |f - p_0|$, $\max |f - p_1|$, and $\max |f - p_2|$ over the interval $[-2, 2]$. Describe how p_2 fits f better around $x = 0$ than does p_1. Why?

b) Let $g(x) = \cos x + x^3$. Verify by hand that g has a minimum at $x = 1$. Calculate by hand the Taylor polynomials $p_0(x)$, $p_1(x)$, and $p_2(x)$ at $x = 1$ of g. Graph these Taylor polynomials together with g on the interval $[-1, 3]$. Compute $E_{max} = \max |g(x) - p_2(x)|$ on intervals $|x - 1| \leq h$ for $h = 1, 1/2, 1/4$, and $1/8$. Each time h is reduced by a factor of $1/2$. By what factor is E_{max} reduced each time?

8. Let $f(x, y) = x \exp(-x^2/2)/(1 + y^2)$.

a) Solve system (8.1) by hand, and verify that f has critical points at $(x, y) = (\pm 1, 0)$.

b) Apply the second-derivative test to determine the nature of each critical point.

c) Using Eq. (8.6), calculate by hand the quadratic approximation $f(1, 0) + q(x, y)$ to f at the critical point $(1, 0)$. Write an mfile for the absolute value of the error, $|E(x, y)| = |f(x, y) - f(1, 0) - q(x, y)|$. Use the mfile `findcrit` on $|E|$ over the square $[0, 2] \times [-1, 1]$. Click down to the square $[1/2, 3/2] \times [-1/2, 1/2]$ and find the maximum of $|E|$ over this smaller square. Do this several times, reducing the side of the square by a factor of 2. Record the max of $|E|$ over each square. By what factor is the maximum error decreased? Does this agree with Eq. (8.7)?

✷   **9.** *Harmonic* functions $f(x, y)$ satisfy the Laplace equation, $f_{xx} + f_{yy} = 0$. We can see immediately that if f is harmonic, then at any critical point, $D \leq 0$. Thus we can never use the second-derivative test to find a local maximum or minimum in the interior of any region. In fact, it can be shown that harmonic functions enjoy a *maximum principle*, which says that the maximum and minimum of a harmonic function over any closed, bounded set K is always attained on the boundary of K.

Let $f(x, y) = \sin x \cosh y$.

a) Verify by hand that f is a harmonic function.

b) Find the critical points of f by solving system (8.1). Using the second-derivative test, verify that each critical point is a saddle point.

c) Use `findcrit` over the starting square $-2\pi \leq x, y \leq 2\pi$. Find the maximum and minimum of f over smaller rectangles. The MATLAB function for $\cosh(x)$ is `cosh`. Where do the maximum and minimum always occur?

10. Suppose point charges of 1 coulomb each are placed at the points $p_1 = (1, 0)$, $p_2 = (-1, 0)$, and $p_3 = (0, h)$. The two-dimensional electrostatic potential from each charge is

$$V_j(x, y) = -\frac{\log(r_j)}{2\pi},$$

where $r_j = \|p - p_j\|$ is the distance from the point p_j. The combined potential is $V = V_1 + V_2 + V_3$.

a) Verify by hand that V is a harmonic function for $p = (x, y) \neq p_j$.

b) Let $h = \sqrt{3}$. Let K be the triangle with vertices p_1, p_2, p_3. Do you think there should be a point in K where V has a minimum? Make an mfile for V, and graph V over the rectangle $\{-1 \leq x \leq 1, .1 \leq y \leq 1.6\}$. There is a critical point at the centroid of the triangle. Use `findcrit` to take a closer look. This is the more complicated kind of behavior that can happen in the case where the discriminant $D = 0$.

c) Now take $h = 2$. You will see that the single critical point at the centroid has split into two critical points. Use `findcrit` to determine their nature.

11. Let $f(x, y) = x^2 - y^2 + 3xy$ and let the set $K = \{(x, y) : x^2 + 2y^2 \leq 1\}$.

a) Use the mfile `lagrange.m` to estimate where the level curves of f are tangent to the boundary of K. Approximately where does the maximum of f over K occur, and what is it? Where does the minimum of f over K occur, and what is it?

b) Solve the Lagrange system (8.9)–(8.11) symbolically, as in Example 8.5. To get a double-precision evaluation of the roots, use `double([lambda,x,y])`.

12. Let $f(x, y, z) = x^2 + 2y^2 + z^2 + xy + x^2y$.

a) Use the second-derivative test to determine the nature of the critical point of f at $(0, 0, 0)$.

b) Use the mfile `impl` over the region $-2 \leq x, y, z \leq 2$. Start with $c = 3$, and decrease c until you see the level set separate into two components. What does the component containing the origin look like for small c?

★   **13.** Let $f(x, y, z) = (xy + z) \exp(-x^2 - y^2 - z^2)$.

a) Verify that f has critical points at $(0, 0, \pm 1/\sqrt{2})$.

b) Use the mfile `impl` on the region $\{-.5 \leq x, y \leq .5, .5 \leq z \leq 1\}$. Letting $c_* = f(0, 0, 1/\sqrt{2})$, display several level sets S_c for values of c near c_*. What is the nature of the critical point at $(0, 0, 1/\sqrt{2})$?

c) Use the second-derivative test to confirm your conclusion in part b).

d) For an interesting view of the level sets for $c = \pm.3$, use `impl` on the region $-1 \leq x, y, z \leq 1$. Combine the two pictures with `hold on`. Rotate around to get a clear view.

★★   **14.** The gravitational potential of a body of mass M located at the origin in three-dimensional space is

$$V(x, y, z) = -\frac{GM}{r}, \qquad r = \sqrt{x^2 + y^2 + z^2}.$$

Here G is the universal gravitational constant. If we have one body of mass M_0 located at the origin and another body of mass M_1 located at the point $(d, 0, 0)$ on the x axis, the combined potential is

$$V(x, y, z) = -\frac{GM_0}{r_0} - \frac{GM_1}{r_1},$$

where $r_0 = \sqrt{x^2 + y^2 + z^2}$ and $r_1 = \sqrt{(x - d)^2 + y^2 + z^2}$.

a) Let $M_0 = 10$, $M_1 = 1$, $G = 1$, and $d = 100$. Graph $V(x, y, 0)$ over the rectangle $R = \{-10 \leq x \leq 110, \ -50 \leq y \leq 50\}$. V tends to $-\infty$ at $(0, 0, 0)$ and at $(d, 0, 0)$, so we must cut off the graph from below. If we want to graph only where $V \geq -2$, we can cut if off with the characteristic function for these points and set $V = -2$ otherwise. Use the following instructions (see Section 5.9):

```
x = linspace(-10, 110, 51); y = linspace(-50, 50, 51);
[X,Y] = meshgrid(x,y);
pot = inline('-10./sqrt(x.^2 +y.^2) -
             1./sqrt((x-100).^2 +y.^2)', 'x', 'y')
V = pot(X,Y);
W = (V+2 >= 0).*(V+2) - 2;
surf(X,Y,W)
```

b) Look for a saddle point of $V(x, y, 0)$ on the x axis between 0 and d. Use the mfile `findcrit` to find an approximate location of the saddle point. Finally, find the saddle point by hand by solving the equation $V_x(x, 0, 0) = 0$.

c) The potential V is rotationally symmetric about the x axis. Rotate the contours of $V(x, y, 0)$ about the x axis to generate the surfaces of gravitational equipotential.

d) If a particle is placed at the saddle point on the x axis, will it fall toward M_0, fall toward M_1, or remain fixed? What if the particle is placed closer to M_0?

★ **15.** Constrained optimization problems often arise in economics. However, the economist is not so interested in finding the maximum or minimum as in observing what happens to the maximum or minimum as certain parameters in the problem change. The following example of this kind of study is called the "Averch Johnson effect."

Recall Example 5.9. There we had a revenue function $R(z) = z^2/(1 + z^2)$ with the Cobb–Douglas production function $z = x^{1/2}y^{1/2}$. Here x is capital and y is labor. The profit function was $\pi(x, y) = R(x, y) - .04x - .06y$.

a) Use findcrit to locate a maximum of π near the point $(3, 2)$.

b) The *rate of return* is the fraction

$$h(x, y) \equiv \frac{R(x, y) - .06y}{x}.$$

What is the rate of return when the profit is maximized as in part a)?

c) Government regulation may be imposed to restrict the rate of return. In mathematical terms, this means we impose the constraint $h(x, y) \leq s$, where s is the permissible rate of return. A smaller value of s corresponds to a more restrictive regulation. Let $G_s = \{(x, y) : h(x, y) < s\}$ be the set of values of capital and labor that yield a permissible rate of return. Let R be the rectangle $\{1 \leq x \leq 6, \ 1 \leq y \leq 3\}$. Use the command contour to plot the level curves $h - s$ for $s - .15, .2, .3, .4$ in R. On which side of the curve does G_s lie? For what values of s does G_s contain the point (x_*, y_*) where the profit was maximized in part a)?

d) The firm now wishes to maximize its profit, subject to the constraint $h \leq s$. It can be shown that the maximum occurs on the curve $h = s$. The constrained maximum problem for the firm is therefore

$$\text{maximize} \quad \pi$$

$$\text{subject to} \quad h = s.$$

The Lagrange equations are

$$\pi_x = \lambda h_x$$

$$\pi_y = \lambda h_y$$

$$h = s. \tag{8.13}$$

Since $h_y = \pi_y/x$, the second equation is simply $(1 - \lambda/x)\pi_y = 0$. It can be shown that $0 < \lambda/x < 1$, so the equations that determine the location of the solution of

(8.13) are

$$\pi_y = 0$$

$$h = s.$$

Use the command `contour` to graph the curve $\pi_y = 0$. Then use `hold` on and the `contour` command again to graph the curves $h = s$ for $s = .4, .3, .2, .15$. Which way does the point of intersection move as s decreases? In particular, what happens to the values of capital and labor as s decreases?

⋆⋆   **16.** The subject of this exercise is usually treated in a course on differential equations. However, with our graphing techniques, we can make some important observations.

Let x and y be two species competing for resources in some environment. $x(t)$ and $y(t)$ will denote the populations of species x and y at time t. The populations change with time, and their rates of change are coupled together in a system of ordinary differential equations,

$$\frac{dx}{dt} = x(a - by)$$

$$\frac{dy}{dt} = y(c - dx). \qquad (8.14)$$

a, b, c, d are positive constants. The growth rate of species x in the absence of competition ($y = 0$) is a; in this case, x grows exponentially, $x(t) = C \exp(at)$. c is the growth rate of y in the absence of competition. The constants b and d describe the interaction of the species (they may like to eat each other). Let $h(x, y) = f(x)g(y)$ where

$$f(x) = x^{-c}e^{dx} \quad \text{and} \quad g(y) = y^a e^{-by}.$$

a) Verify by hand that if $(x(t), y(t))$ is a solution pair of system (8.14), then $(dh/dt)(x(t), y(t)) = 0$. This means that the solution curves $t \to (x(t), y(t))$ lie on the level curves of h.

b) Verify that $f'(a/b) = g'(c/d) = 0$ and $f''(c/d) > 0$ while $g''(a/b) < 0$.

c) Show that part b) implies that $(c/d, a/b)$ is critical point of h that is a saddle point. Note further that $x(t) \equiv c/d$ and $y(t) \equiv a/b$ is a solution pair of (8.14). Thus the critical point $(c/d, a/b)$ is a state of the system in which the two species are in perfect equilibrium.

d) Now make an mfile `h.m` for h as a function $h(x, y, p)$, where $p = [a, b, c, d]$ is a vector of the coefficients. Make a fine mesh (100×100) over the rectangle $R = [0, 10] \times [0, 4]$. Start with $p = [a, b, c, d] = [.5, .4, .6, .3]$. Plot the level curves of h over R using 40 levels. Calculate the value $h_0 = h(c/d, a/b, p)$ for this choice of p. Use the command `contour(X,Y,h(X,Y,p), [h0,h0], 'r')` to add the level set $h = h_0$ that passes through the critical point. This level set consists of two curves that intersect at the critical point. You should see the typical curves of a saddle point.

e) The solution curves $(x(t), y(t))$ of (8.14) provide a parameterization of the level curves of h. Use the MATLAB function `quiver` to add the vector field $(u, v) = (x(b - ay), y(c - dx))$ to the plot of the level curves you made in part d). Use a much coarser mesh (10×10) with the `quiver` function.

f) If at time $t = 0$, the system is in a state (x_0, y_0) lying on one of the level curves of h, the state of the system will evolve along the level curve in the direction indicated by the arrows. If the level curve tends toward either of the axes, one of the species is dying out. Let Γ be the part of the level set $h = h_0$ that is the curve with positive slope. If the initial state (x_0, y_0) lies above and to the left of Γ, which species will die out? What happens if (x_0, y_0) lies below and to the right of Γ? What happens if (x_0, y_0) lies on Γ? What is the realistic probability that the system tends toward equilibrium?

g) Now set $a = b = c = 1$, and let d vary from $.1$ to 1. When d is small, the x species has a small effect on the growth of the y species. Explain in terms of the species interaction why the equilibrium point $(1/d, 1)$ moves to the right as d decreases.

h) The predator–prey relationship is modeled (with x the prey and y the predator) by the following system of differential equations:

$$\frac{dx}{dt} = ax - bxy$$

$$\frac{dy}{dt} = -cy + dxy.$$

Continue to assume that a, b, c, d are positive constants. The solution curves of this system lie on the level curves of the function $k(x, y) = g(y)/f(x)$. The point $(c/d, b/a)$ is again an equilibrium point for this system and a critical point for k. Investigate the level curves of k, add the appropriate vector field, and interpret the results in terms of the predator–prey relationship.

9

Multiple Integrals

Prepared mfiles used in this chapter

```
riemann   simp2   simp3   trf
```

9.1   Double integrals over rectangles

The double integral of a function $f(x, y)$ over a rectangle R is usually approached by placing a grid over the rectangle that divides R into subrectangles $R_{i,j}$. We then pick a point $p_{i,j}$ in each subrectangle $R_{i,j}$ and approximate f on each $R_{i,j}$ by the constant value $f(p_{i,j})$. The Riemann sum corresponding to this subdivision and to this choice of $p_{i,j}$ is

$$S(f) = \sum_{i,j} f(p_{i,j}) A(R_{i,j}) \tag{9.1}$$

where $A(R_{i,j})$ is the area of each subrectangle. When $f(x, y)$ is a continuous function on R, these Riemann sums converge to a limit as the subdivisions become finer and finer. This limit is defined to be the double integral of f over R, and we write

$$\int \int_R f(x, y) \, dA = \lim_{\Delta \to 0} S(f).$$

Here Δ is the maximum diagonal over all the subrectangles $R_{i,j}$.

mfile riemann

The mfile `riemann.m` computes Riemann sums and provides a graphical display of the approximating piecewise constant function. The call is `riemann(f, corners)`,

169

where f is a function of two variables given as an inline function. When f is given in an mfile, the call is `riemann('f', corners)`. As usual, `corners` is a vector $[a, b, c, d]$, where the rectangle R is defined as $R = \{(x, y) : a \leq x \leq b, \; c \leq y \leq d\}$. We will divide the interval $[a, b]$ into n subintervals, and the interval $[c, d]$ into m subintervals. When the call is made, the user chooses m and n and the Riemann sum is computed, using as a choice for $p_{i,j}$ the center of each subrectangle $R_{i,j}$. Here are the sequence of commands in `riemann.m` after a, b, c, d, m, and n are entered:

```
dx = (b-a)/n;   dy = (d-c)/m;
% calculate the center points of the subrectangles
p = a +.5*dx: dx : b-.5*dx;
q = c +.5*dy: dy : d-.5*dy;
[P,Q] = meshgrid(p,q);
Z = f(P,Q);

disp(' Approximate value of the integral ')
integral = sum(sum(Z))*dx*dy;
```

The Riemann sum is computed in the last line using the command `sum`. When `sum` is applied to a matrix, the sum of each column is computed and stored in a row vector. When `sum` is applied to a vector, row, or column, the elements of the vector are summed.

A third argument may be added to the call for `riemann`. When the call is `riemann(f, corners, graph)`, a graph of the approximating step function is displayed. `graph` can be any number, such as 1 or 2.

Example 9.1

Let $f(x, y) = x^3 + x + y$ and let $R = \{1 \leq x \leq 2, \; 0 \leq y \leq 3\}$.

```
>> f = inline('x.^3 + x + y', 'x', 'y')
>> corners = [1 2 0 3];
>> riemann(f, corners, 1)
enter the number of subdivisions in x and y
                     directions as [n m]   [10 20]
Approximate value of the integral
ans =
     20.2387
```

The result is displayed in Figure 9.1. The limiting value of the Riemann sums is $\int\int_R (x^3 + x + y)\, dx\, dy = 20.25$.

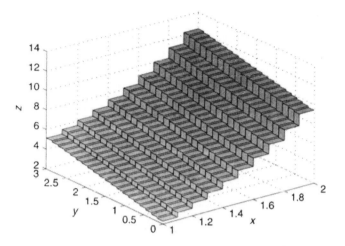

Figure 9.1 Step function approximation to the graph of $f(x, y) = x^3 + x + y$ over the rectangle $R = \{(x, y) : 1 \le x \le 2, \ 0 \le y \le 3\}$.

Symbolic Integration

Double integrals over a rectangle can be expressed as iterated integrals,

$$\int\int f(x, y)\, dA = \int_a^b \int_c^d f(x, y)\, dy\, dx = \int_c^d \int_a^b f(x, y)\, dx\, dy.$$

In this form we can use the symbolic integrator of MATLAB to compute first the inner integral and then the outer integral.

Recall from Section 1.6 that the command for the symbolic integrator is int(g,a,b) when $g = g(x)$ is a function of one variable. When we have a function of two variables, such as $f(x, y)$, the integration command must specify which is the variable of integration. Thus if we wish to integrate with respect to y from c to d, the command is G = int(f,y,c,d). This will produce a symbolic function $G(x) = \int_c^d f(x, y)\, dy$, which can now be integrated from a to b with int(G,a,b). The two commands can be combined as int(int(f,y,c,d),a,b).

Example 9.2

Let $f(x, y) = x^3 + x + y$ as in Example 9.1.

```
>> syms x y
>> f = x^3 +x+y
>> int(int(f,y,0,3), 1,2)
ans =
81/4
```

If we wish to keep the limits of integration as parameters in the answer, we declare a, b, c, d as symbolic variables and then repeat the preceding sequence of commands:

```
>> syms x y a b c d
>> f = x^3+x+y
>> int(int(f,y,c,d),a,b)
ans =
  1/4*b^4*d+1/2*b^2*d+1/2*d^2*b-1/4*b^4*c
  -1/2*b^2*c-1/2*c^2*b-1/4*a^4*d- 1/2*a^2*d
  -1/2*d^2*a+1/4*a^4*c+1/2*a^2*c+1/2*c^2*a
```

Numerical methods

The numerical integration scheme used in the mfile `riemann.m` is a two-dimensional analog of the one-dimensional midpoint rule. It is natural to look for two-dimensional analogs of other rules, such as Simpson's rule. Recall that for Simpson's rule in one dimension, we subdivide the interval $[a, b]$ into n equal subintervals of length $h = (b - a)/n$, with n even. Then

$$\int_a^b g(x)\, dx = S_n(g) + E_n(g)$$

where

$$S_n(g) = \frac{h}{3}[g(x_1) + 4g(x_2) + 2g(x_3) + \cdots + 2g(x_{n-1}) + 4g(x_n) + g(x_{n+1})] \quad (9.2)$$

and the error

$$E_n(g) = -\frac{1}{180}(b - a)h^4 g^{(4)}(\xi) \quad (9.3)$$

for some point ξ in the interval $[a, b]$. Simpson's rule is derived by fitting parabolas through the points $(x_i, g(x_i))$. We would expect Simpson's rule to be exact on quadratic polynomials, but in fact it is exact on cubics, because if g is a cubic, then $g^{(4)} = 0$ and the error term is zero. The presence of h^4 in the error term means that

$$E_{2n}(g) \approx \frac{1}{16} E_n(g).$$

Simpson's rule is very easy to implement in MATLAB. We could use a for loop with summation, but MATLAB is faster when for loops are avoided. We recognize that the rule can be expressed as

$$\left(\frac{h}{3}\right) \mathbf{g} \cdot \mathbf{s},$$

where $\mathbf{s}$ is the *Simpson vector*, $\mathbf{s} = [1, 4, 2, \ldots, 2, 4, 1]$, and $\mathbf{g} = [g(x_1), \ldots, g(x_{n+1})]$. A short script mfile to implement Simpson's rule is

```
x = linspace(a,b,n+1);
svec = 2*ones(1,n+1);
svec(2:2:n) = 4*ones(1,n/2);
svec(1) = 1; svec(n+1) = 1;
integral = dot(g(x),svec)*(b-a)/(3*n);
```

Here we assume that a, b, and n have been entered and that g is given in an mfile or as an inline function that is array-smart. The Simpson vector, `svec`, is constructed by taking a vector that is all 2's and setting the even indexed components to 4. Finally the first and last components are set to 1.

For future use, we write a short function mfile that yields the Simpson vector for n subintervals:

```
function s = simpvec(n)
   s = 2*ones(1,n+1);
   s(2:2:n) = 4*ones(1,n/2);
   s(1) = 1; s(n+1) =1;
```

What is the two-dimensional analog of Simpson's rule? Write the double integral over the rectangle R as an iterated integral,

$$\int \int_R f(x, y) \, dA = \int_c^d F(y) \, dy$$

where

$$F(y) = \int_a^b f(x, y) \, dx.$$

Now divide the interval $[a, b]$ into n (n must be even) equal subintervals, with points $a = x_1 < x_2 < \cdots < x_{n+1} = b$. We denote the Simpson vector in the x direction as

$$\mathbf{s}^x = [1, 4, 2, \ldots, 2, 4, 1] = [s_1^x, s_2^x, \ldots s_n^x, s_{n+1}^x].$$

Then for each y, the Simpson's rule estimate for $F(y)$ is

$$F(y) \approx \frac{b-a}{3n} \sum_{j=1}^{n+1} s_j^x f(x_j, y). \tag{9.4}$$

Similarly, we divide the interval $[c, d]$ into m (m must be even) equal subintervals $c < y_1 < y_2 < \cdots < y_{m+1} = d$. The Simpson vector in the y direction is

$$\mathbf{s}^y = [1, 4, 2, \ldots, 2, 4, 1] = [s_1^y, s_2^y, \ldots s_{m+1}^y].$$

Then the Simpson's rule estimate for the integral of F is

$$\int_c^d F(y)\,dy \approx \frac{d-c}{3m} \sum_{i=1}^{m+1} s_i^y F(y_i). \tag{9.5}$$

For each i, we replace $F(y_i)$ in the right side of Eq. (9.5) with the Simpson estimate (9.4). Making this replacement, we have

$$\int\int_R f(x,y)\,dA = \int_c^d F(y)\,dy \approx \frac{(b-a)(d-c)}{9mn} \sum_{i,j} s_j^x s_i^y f(x_j, y_i).$$

The $(m+1) \times (n+1)$ matrix S with $S_{i,j} = s_j^x s_i^y$ is the *Simpson matrix*.

The two-dimensional version of Simpson's rule is also quite easy to implement in MATLAB using the array operations. Assuming a, b, c, d, m, and n have been entered and that $f(x, y)$ is given in an mfile or as an inline function, the script is:

```
x   =   linspace(a,b,n+1);
y   =   linspace(c,d,m+1);
[X,Y]   =   meshgrid(x,y);

% Construct the Simpson vector in x.
svecx = simpvec(n);

% Construct the Simpson vector in y
svecy = simpvec(m);

% Construct the Simpson matrix
S = svecy'*svecx;
integral = sum(sum(S.*f(X,Y)))*(b-a)*(d-c)/(9*m*n);
```

The Simpson matrix is constructed with the operation $S = svecy'.*svecx$, which is the product of the $(m+1) \times 1$ vector $svecy'$ and the $1 \times (n+1)$ vector $svecx$. $f(X,Y)$ is the $(m+1) \times (n+1)$ matrix with elements $f(x_j, y_i)$. The Simpson matrix for $n = 6, m = 4$ is the 5×7 matrix

$$S = \begin{bmatrix} 1 & 4 & 2 & 4 & 2 & 4 & 1 \\ 4 & 16 & 8 & 16 & 8 & 16 & 4 \\ 2 & 8 & 4 & 8 & 4 & 8 & 2 \\ 4 & 16 & 8 & 16 & 8 & 16 & 4 \\ 1 & 4 & 2 & 4 & 2 & 4 & 1 \end{bmatrix}.$$

mfile simp2.m

The mfile `simp2.m` is a function mfile with the call `simp2(f,corners)` when f is an inline function and `simp2('f',corners)` when f is given in an mfile. As in the mfile `riemann.m`, `corners` is the vector $[a, b, c, d]$ that determines the rectangle $R = \{a \le x \le b, \ c \le y \le d\}$. After the call the program pauses to let you enter the number of subdivisions n in the x direction and m in the y direction.

Example 9.3

Let $f(x, y) = xy^2 + \cos(x + y)^2$ and the rectangle $R = [1, 2] \times [0, 3]$.

```
>> f = inline('x.*y.^2 + cos(x+y).^2', 'x', 'y')
>> corners = [1 2 0 3]
>> s1 = simp2(f, corners)
enter the number of subdivisions in
              x and y directions as [n m]   [10 20]
Approximate value of the
              integral using Simpson's rule
s1 =
15.05701244123409
```

The numerical approximation produced by any numerical method is useless and can be quite misleading if we do not have some estimate of the error in this approximation. It may seem impossible to obtain such an estimate since we usually do not know the exact result. However, in the case of Simpson's rule we can make an error estimate easily. Double the number of subdivisions in each direction, and compute the approximation of the integral again. The difference in the two numerical calculations is an estimate of the error made in the first calculation. This estimate works remarkably well, even for functions where the derivatives have singularities, e.g., $f(x) = \sqrt{x}$. However, it is not to be trusted when f itself has singularities.

In the preceding example, we apply Simpson's rule again with $n = 20$ and $m = 40$. The result is

```
>> s2 = simp2(f, corners)
enter the number of subdivisions in
              x and y directions as [n,m]   [20 40]
Approximate value of the
              integral using Simpson's rule
s2 =
15.05700953134787

>> s1 - s2
=  2.909886214652602e-06
```

Hence, the error in the first calculation is on the order of 3×10^{-6}.

MATLAB integrators

MATLAB has its own built-in integration routines. The one-dimensional routines are quad and quadl. They use methods similar to Simpson's rule but apply them in a more sophisticated manner called *adaptive quadrature*. From Eq. (9.3) for the error in Simpson's rule, we see that a naive, straightforward way to reduce the error is to refine the mesh uniformly over the interval $[a, b]$ by cutting h in half, i.e., by doubling n. However, quad and quadl are more efficient, in that they refine the mesh only where the function is changing more rapidly. The calls for quad and quadl are quad(f,a,b) and quadl(f,a,b) when f is given as an inline function and use single quotes when f is given in an mfile. The interval of integration is $[a, b]$. These integrators estimate the error in their results and, unless specified otherwise, continue to refine the mesh adaptively until the relative error is less than 10^{-3}. A different tolerance for the relative error may be specified by adding a fourth argument to the call, quad(f,a,b,tol), where tol is the desired tolerance. For example, we could take tol = 10^(-4).

It is possible to iterate these routines to evaluate double integrals, but there are some delicate points related to passing the variables of integration to the functions involved. Probably for this reason, there is a built-in double integrator, dblquad. The call for dblquad is

$$\text{dblquad}(f,a,b,c,d)$$

Here it is assumed that the "inner" integration is done in x, $a \le x \le b$, and that the second, "outer" integration is done in y, $c \le y \le d$ for $f(x, y)$. If it is desired to do the integration in y first, the call will be

$$\text{dblquad}(g,c,d,a,b)$$

where $g(y, x) = f(x, y)$. Note the reversal of x, y.

We have introduced our own double integrator simp2 for two reasons. First, Simpson's rule is easy to understand in several variables. Second, dblquad appears to find possible singularities in the integrand when there are none, and sometimes exits before achieving the desired error tolerance.

Example 9.4

We shall integrate the function $f(x, y) = (4/\pi) \exp(-x^2 - y^2)$ over the rectangle $R = \{0 \le x \le .5, 0 \le y \le 1\}$. The result can be expressed in terms of the error function:

$$\text{erf}(x) = \frac{2}{\sqrt{\pi}} \int_0^x e^{-s^2} ds.$$

We have

$$\int\int_R f(x,y)\,dA(x,y)\frac{4}{\pi}\int_0^{.5}e^{-x^2}\,dx\int_0^1 e^{-y^2}\,dy = \text{erf}(.5)\text{erf}(1).$$

The following script estimates the integral, first using dblquad and then using simp2. Finally, a more accurate estimate of the integral is obtained by using the MATLAB routine for evaluating the error function, erf:

```
>> f = inline('exp(-x.^2 -y.^2)*(4/pi)', 'x', 'y')
>> dblquad(f,0,.5, 0,1)
ans =
0.43862714228015

>>   corners = [0 0.5 0 1]
>> simp2(f,corners)
enter the number of subdivisions [n m] [10 20]
ans =
0.43862581863650

>> erf(.5)*erf(1)
ans =
0.43862565976328
```

9.2   Nonrectangular regions of integration

The double integral over a rectangle was defined as the limit of the Riemann sums (9.1), when this limit exists. How can this concept be adapted to a set G that is not a rectangle?

Let $f(x, y)$ be a function that we wish to integrate over the set G. We define the restriction of f to G as

$$\tilde{f}(x, y) = \begin{cases} f(x, y) & \text{for } (x, y) \in G \\ 0 & \text{for } (x, y) \notin G \end{cases}.$$

Now let R be a rectangle that contains G. For example, in Figure 9.2, the rectangle $R = \{-1.5 \le x \le 1.5,\ -1 \le y \le .5\}$. Compute the Riemann sums of $\tilde{f}$ over R. If the boundary of G is piecewise smooth and f is continuous on G, it can be shown that these Riemann sums converge, and we define the double integral $\int\int_G f\,dx\,dy$ as the limit of these sums. See Figure 9.2.

Because of the discontinuity of the function at the boundary of G, the direct Riemann sum approach will require a very fine mesh on the boundary to get an

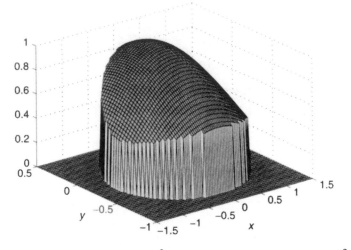

Figure 9.2 Graph of $f(x, y) = \cos(x) \exp(-y^2)$ restricted to the ellipse $G = \{x^2 + 2y + 4y^2 \le 1\}$. G is contained in the rectangle $R = \{|x| \le 1.5, \ -1 \le y \le .5\}$.

accurate result. An adaptive two-dimensional method would be needed that would use a fine mesh on the boundary and a coarser mesh in the interior of G. This can be difficult to implement.

In some cases, the geometry of a nonrectangular region G allows $\int \int_G f \, dA$ to be computed as an iterated integral. For instance, if G is vertically simple, $G = \{(x, y) : c(x) \le y \le d(x), \ a \le x \le b\}$, we can write

$$\int \int_G f \, dA = \int_a^b \int_{c(x)}^{d(x)} f(x, y) \, dy \, dx.$$

We may be able to use the symbolic integrators to calculate such an integral.

Example 9.5

The set G in Figure 9.3 is described as

$$G = \{(x, y) : x(3 - x) \le y \le \sin(x), 0 \le x \le 2.4\}.$$

Let $f(x, y) = xy$. We can compute $\int \int_G f(x, y) \, dA$ as follows:

```
>> syms x y
>> f = x*y
>> F = int(f,y,x*(x-3), sin(x))

F =
```

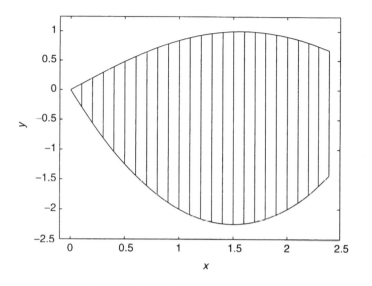

Figure 9.3 Vertically simple region.

```
1/2*sin(u)^2*u  1/8*u^5/3*u^1  9/2*x^5

>> int(F,0,2.4)
ans =
-3/5*cos(12/5)*sin(12/5)-74286/15625+1/8*sin(12/5)^2
>> double(ans)
ans =
-4.3894
```

In many cases double integrals cannot be done symbolically, and we must use numerical methods. The iterated form of the integral over a nonrectangular region is often not the best way to make a numerical estimate of the integral. A better way is to make a change of coordinates so that the set becomes a rectangle. Then we can apply the numerical methods developed for rectangles. We must postpone this treatment until we determine how the area of a set is affected by a change of variable.

9.3   Change of variable in double integrals

Affine transformations of the plane

A pair of functions $x = f(u, v)$ and $y = g(u, v)$ takes a point with coordinates (u, v) and maps it into the point with coordinates $(x, y) = (f(u, v), g(u, v))$. We can think of this operation as a mapping of one copy of the plane, with coordinates

u, v, into or onto another copy of the plane, with coordinates x, y. The mapping $(x, y) = T(u, v) = (f(u, v), g(u, v))$ is called a *transformation*. For example, $T(u, v) = (uv, u + v)$.

An important class of these transformations are the *affine* transformations. They have the form

$$x = f(u, v) = Au + Bv + w_1 \qquad (9.6)$$
$$y = g(u, v) = Cu + Dv + w_2.$$

We write the transformation more compactly as

$$(x, y) = T(u, v)$$

and note that $T(0, 0) = w = (w_1, w_2)$. When $(w_1, w_2) = (0, 0)$, T takes $(0, 0)$ into $(0, 0)$ and is said to be a *linear* transformation.

It is not hard to show that an affine transformation takes a parallelogram into a parallelogram. Furthermore, there is a simple formula that describes how the area is changed. If Q is a parallelogram in u, v space and P is the image of Q under the affine transformation T, then

$$\text{area}(P) = |AD - BC| \, \text{area}(Q). \qquad (9.7)$$

We shall see later that this rule for the change of area of a parallelogram under an affine transformation works for all sets having area.

General transformations of the plane

Now we ask how a more general transformation changes area. We concentrate on the case of a rectangle $R = \{a \leq x \leq b, \ c \leq y \leq d\}$. The transformation $T(u, v) = (f(u, v), g(u, v))$ will map R into a set G with possibly curvilinear boundaries.

In Figure 9.4 the square $R = [1, 2] \times [0, 1]$ is mapped by

$$T(u, v) = (u^2 - v^2, 2uv)$$

into the curvilinear figure on the right.

We can approximate the image set G by a parallelogram P that is the image of R under an affine transformation $\tilde{T}$. The affine transformation $\tilde{T}$ that approximates T is obtained by using the tangent plane approximations for f and g at a point $(u_0, v_0) \in R$:

$$f(u, v) \approx f(u_0, v_0) + f_u(u_0, v_0)(u - u_0) + f_v(u_0, v_0)(v - v_0) \qquad (9.8)$$
$$g(u, v) \approx g(u_0, v_0) + g_u(u_0, v_0)(u - u_0) + g_v(u_0, v_0)(v - v_0). \qquad (9.9)$$

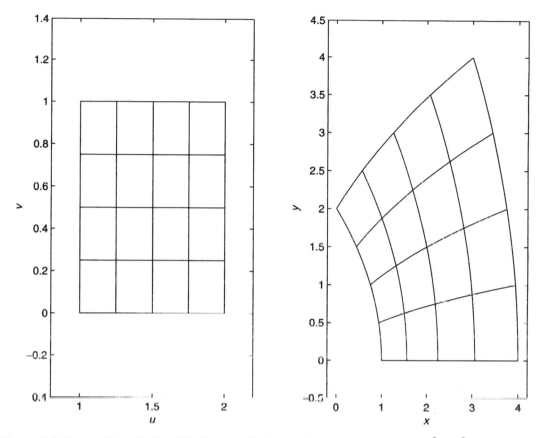

Figure 9.4 Square $R = [1, 2] \times [0, 1]$ on the left, transformed by $T(u, v) = (u^2 - v^2, 2uv)$. Image set G is on the right.

Recall from Chapter 7 that the Jacobian matrix of a pair of functions $f(u, v)$ and $g(u, v)$ is

$$J(u, v) = \begin{bmatrix} f_u(u, v) & f_v(u, v) \\ g_u(u, v) & g_v(u, v) \end{bmatrix}.$$

Approximations (9.8) and (9.9) can be written

$$T(u, v) = (f(u, v), g(u, v)) \approx T(u_0, v_0) + J(u_0, v_0) \begin{bmatrix} u - u_0 \\ v - v_0 \end{bmatrix}.$$

In the right-hand side the matrix product,

$$J(u_0, v_0) \begin{bmatrix} u - u_0 \\ v - v_0 \end{bmatrix} = \begin{bmatrix} f_u(u_0, v_0)(u - u_0) + f_v(u_0, v_0)(v - v_0) \\ g_u(u_0, v_0)(u - u_0) + g_v(u_0, v_0)(v - v_0) \end{bmatrix}.$$

The approximating affine transformation, with $A = f_u(u_0, v_0)$, $B = f_v(u_0, v_0)$, $C = g_u(u_0, v_0)$, $D = g_v(u_0, v_0)$, is therefore

$$\tilde{T}(u, v) = T(u_0, v_0) + J(u_0, v_0) \begin{bmatrix} u - u_0 \\ v - v_0 \end{bmatrix}. \tag{9.10}$$

The point (u_0, v_0) may be taken anywhere in the rectangle R. We shall make the choice of the midpoint $(u_0, v_0) = ((a + b)/2, (c + d)/2)$. The multiplication factor of the area of the affine approximation is

$$|AD - BC| = |\det J((a + b)/2, (c + d)/2)|.$$

We shall abbreviate this determinant expression by writing simply $|J(u, v)|$ for $|\det J(u, v)|$.

Thus using Eq. (9.7), the area of the parallelogram P, which is the image of the rectangle R under $\tilde{T}$, is

$$\text{area}(P) = |J((a + b)/2, (c + d)/2)|(b - a)(d - c).$$

To get a better approximation of the area of G, we subdivide the rectangle R into subrectangles $R_{i,j}$ and make an affine approximation $\tilde{T}_{i,j}$ on each $R_{i,j}$. These affine approximations produce parallelograms $P_{i,j} = \tilde{T}_{i,j}(R_{i,j})$. As the number of subrectangles increases, the union of the parallelograms $P_{i,j}$ provides a better and better approximation to the image set G. The sum of the areas of the $P_{i,j}$ is also an approximation to the area of G. In fact we have

$$\int\int_G dA(x, y) = A(G) \approx \sum_{i,j} A(P_{i,j})$$

$$= \sum_{i,j} |J(u_i, v_j)| A(R_{i,j})$$

where we have chosen (u_i, v_j) to be the midpoints of the subrectangles $R_{i,j}$. In the limit, as the diameter of the rectangles $R_{i,j}$ tends to zero, we obtain

$$\text{area}(G) = \int\int_G dA(x, y) = \int\int_R |J(u, v)| \, dA(u, v). \tag{9.11}$$

mfile trf

The mfile trf.m displays this approximation process. It transforms a rectangle into a curvilinear set G and makes an approximation to the area of G by adding the areas of the approximating parallelograms. The user must provide functions $f(u, v)$ and $g(u, v)$ in mfiles or as inline functions. These functions define the transformation. A rectangle R is defined by the usual vector corners. The call is trf(f,g,corners) when f and g are given as inline functions and trf('f', 'g', corners) when f and g are given in mfiles. The user must then enter the number of subdivisions in each direction. The rectangle R is displayed on the left of the figure, and the curvilinear set G is displayed on the right with the images of the subrectangles. To see the approximating parallelograms superimposed on G, the user then hits return. The sum of the areas of the rectangles is displayed on the screen.

Example 9.6

We use the same transformation used to make Figure 9.4,

$$T(u, v) = (f(u, v), g(u, v)) = (u^2 - v^2, 2uv),$$

and the rectangle $R - [1, 2] \times [0, 1]$.

```
>> f = inline('u.^2 - v.^2', 'u', 'v')
>> g = inline('2*u.*v', 'u', 'v')
>> corners = [1 2 0 1]
>> trf(f,g,corners)
enter the number of subdivisions [m n]  [4 4]
(hit return)
A =
     10.625
```

The result is shown in Figure 9.5.

Formula (9.11) extends to the integral of a function $h(x, y)$ over the set G. The general formula becomes

$$\int \int_G h(x, y) \, dA(x, y) = \int \int_R \tilde{h}(u, v) |J(u, v)| \, dA(u, v). \qquad (9.12)$$

Here the function $\tilde{h}(u, v) = h(f(u, v), g(u, v))$ is the composition of the transformation T and the function h. We call $\tilde{h}$ the *pullback* of h to the rectangle R.

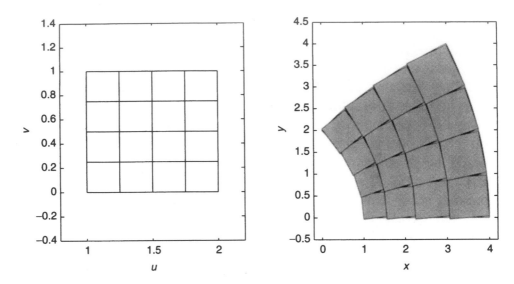

Figure 9.5 Approximating parallelograms produced by affine transformations on each subrectangle, $m = n = 4$.

Change-of-variable formula (9.12) is valid in the more general situation when G is the image under a transformation T of another curvilinear set $\tilde{G}$. In this case we would say that $\tilde{G}$ is the *pullback* of G under the transformation T.

A great advantage of finding a pullback of G that is a rectangle R is that we can apply the numerical means of estimating integrals over rectangles to the integral of $\tilde{h}$ over R.

Vertically simple and horizontally simple regions G

We return to the question of numerical estimation of integrals over vertically or horizontally simple regions. Recall that if G is a vertically simple region, we can describe G as

$$G = \{(x, y): \ c(x) \le y \le d(x), \ a \le x \le b\}.$$

Let R be the rectangle $R = \{(u, v) : a \le u \le b, \ 0 \le v \le 1\}$. The simple transformation

$$T(u, v) = (f(u, v), g(u, v)) = (u, (1 - v)c(u) + vd(u))$$

maps the rectangle R onto the set G. Notice that the lower edge of R, where $v = 0$, is mapped onto the curve $y = c(x)$ and that the upper edge, $v = 1$, is mapped onto the curve $y = d(x)$. The Jacobian matrix of this transformation is

$$J(u, v) = \begin{bmatrix} 1 & 0 \\ (1 - v)c'(u) + vd'(u) & d(u) - c(u) \end{bmatrix}$$

and the Jacobian determinant is

$$|J(u, v)| = |d(u) - c(u)|.$$

Hence the integral over G becomes

$$\int\int_G h(x, y)\, dA\, (x, y) = \int\int_R \tilde{h}(u, v)|d(u) - c(u)|\, dv\, du$$

where $\tilde{h}(u, v) = h(u, (1 - v)c(u) + vd(u))$.

Example 9.7

Let $h(x, y) = \cos^2(x)\exp(-xy)$, and let

$$G - \{(x, y) : x(x - 3) \le y \le \sin(x),\ 0 \le x \le \pi/2\}.$$

Then

$$\int\int_G h(x, y)\, dA(x, y) = \int_0^{\pi/2}\int_0^1 \tilde{h}(u, v)|\sin u - u(u - 3)|\, dv\, du$$

where

$$\begin{aligned} \tilde{h}(u, v) &= h(u, (1 - v)c(u) + vd(u)) \\ &= \cos^2(u)e^{-u((1-v)u(u-3)+v\sin(u))} \end{aligned}$$

We estimate this integral with the mfile `simp2.m`. First we write an mfile for the integrand $q(u, v) \equiv \tilde{h}(u, v)(d(u) - c(u))$:

```
function z = q(u,v)
x = u;
y = (1-v).*u.*(u-3) + v.*sin(u);
z = cos(x).^2 .*exp(-x.*y).*abs(sin(u) - u.*(u-3));
. . . . . . . . . . . . . . . . . . . . .

>> format long
```

```
>> corners = [0 pi/2 0 1]
>> simp2('q',corners)
enter the number of subdivisions [n m] [30 20]

Approximate value of the integral
                          using Simpson's rule
ans =
    2.03220348853753
```

To get an estimate of the error in this result, we run Simpson's rule again, this time doubling the number of subdivisions in each direction:

```
>> simp2('q', corners)
enter the number of subdivisions [n m] [60 40]

Approximate value of the integral
                          using Simpson's rule
ans =
    2.03219215873978
```

The difference in the results is about 10^{-5}, which means that the absolute value of the error in the first result is on the order of 10^{-5}.

A similar change of variables to a rectangle can be made for regions that are horizontally simple.

Polar coordinates

An important special case of the change-of-variable formula is that of polar coordinates. We suppose that a set G in the xy plane is the image under the transformation

$$x = r \cos \theta, \quad y = r \sin \theta$$

of some rectangle R in the r, θ plane. The Jacobian matrix of this transformation is

$$J(r, \theta) = \begin{bmatrix} x_r & x_\theta \\ y_r & y_\theta \end{bmatrix} = \begin{bmatrix} \cos \theta & -r \sin \theta \\ \sin \theta & r \cos \theta \end{bmatrix}$$

and

$$|\det J(r, \theta)| = r((\cos \theta)^2 + (\sin \theta)^2) = r.$$

Thus for a set G, which in polar coordinates is described by $a \le r \le b$ and $c \le \theta \le d$, we have

$$\int\int_G h(x, y)\, dA(x, y) = \int_c^d \int_a^b h(r \cos \theta, r \sin \theta)\, r\, dr\, d\theta.$$

More generally, if G is described in polar coordinates as

$$\tilde{G} = \{(r, \theta) : a(\theta) \le r \le b(\theta), \ c \le \theta \le d\},$$

then

$$\int\int_G h(x, y)\, dA(x, y) = \int\int_{\tilde{G}} h(r\cos\theta, r\sin\theta)\, r\, dr\, d\theta.$$

This integral may in turn be pulled back to the rectangle $R = \{c \le u \le d, \ 0 \le v \le 1\}$ with the change of variable

$$r = r(u, v) = (1 - v)a(u) + vb(u), \quad \theta = u.$$

This yields

$$\int\int_G h(x, y)\, dA(x, y) = \int\int_R \tilde{h}(u, v)r(u, v)|b(u) - a(u)|\, dA(u, v)$$

where

$$\tilde{h}(u, v) = h(r(u, v)\cos u, r(u, v)\sin u).$$

Example 9.8

The set G bounded by a cardiod curve is

$$G = \{(r, \theta) : 0 \le r \le 1 + \sin\theta, \ 0 \le \theta \le 2\pi\}.$$

Let $h(x, y) = \exp(-x^2 + y)$. First we put in polar coordinates. We have

$$\int\int_G h(x, y)\, dA(x, y) = \int_0^{2\pi} \int_0^{1+\sin\theta} h(r\cos\theta, r\sin\theta)\, r\, dr\, d\theta.$$

Then we make a second change of variable,

$$r = r(u, v) = v(1 + \sin u), \quad \theta = u.$$

where $0 \le u \le 2\pi$, and $0 \le v \le 1$. Finally, we have

$$\int\int_G h(x, y)\, dA(x, y) = \int_0^1 \int_0^{2\pi} \tilde{h}(u, v)r(u, v)(1 + \sin u)\, du\, dv.$$

The double pullback $\tilde{h}(u, v)$ is defined by the compositions

$$\tilde{h} = \exp(-x^2 + y) = \exp(-(r(u, v)\cos u)^2 + r(u, v)\sin u).$$

We estimate this integral using the mfile `simp2`. Let

$$q(u, v) = \tilde{h}(u, v)r(u, v)(1 + \sin u).$$

```
function z = q(u,v)
    r = v.*(1 + sin(u));
    x = r.*cos(u); y = r.*sin(u);
    z = exp(-x.^2 + y).*r.*(1 + sin(u));
```
. . . .

```
>> corners = [0 2*pi 0 1]
>> simp2('q', corners)
enter the number of subdivisions [n m]  [120 20]
ans =
9.17859917245639

>> simp2('q', corners)
enter the number of subdivisions [n m]  [240 40]
ans =
9.17857577346632
```

When m and n are doubled, the results differ by 2.3×10^{-5}.

9.4   Triple integrals

The integral $\int \int \int_R f \, dV$, where now R is a rectangular solid in x, y, z space, is defined in the same way as the limit of Riemann sums. These triple integrals can be represented as iterated integrals. Suppose the solid $R = \{(x, y, z) : a_1 \leq x \leq a_2, b_1 \leq y \leq b_2, c_1 \leq z \leq c_2\}$. Then

$$
\int \int \int_R f(x, y, z) \, dV = \int_{a_1}^{a_2} \int_{b_1}^{b_2} \int_{c_1}^{c_2} f(x, y, z) \, dz \, dy \, dx
$$

$$
= \int_{b_1}^{b_2} \int_{a_1}^{a_2} \int_{c_1}^{c_2} f(x, y, z) \, dz \, dx \, dy
$$

$$
= \int_{c_1}^{c_2} \int_{b_1}^{b_2} \int_{a_1}^{a_2} f(x, y, z) \, dx \, dy \, dz
$$

Altogether there are six different orders in which we may do the triple integral. Once written as an iterated integral, we may apply the symbolic integrator `int` three times and hope that the integral can be done this way. If not, we can use a three-dimensional version of Simpson's rule, which is constructed in the same way as the two-dimensional rule. It is implemented in the mfile `simp3`.

mfile simp3.m

The call is `simp3(f,corners)`, where f is given in an mfile or as an inline function. Here, `corners` is a six-vector of the corner coordinates of the rectangular solid. After the call, the user is asked to specify the number or subdivisions in each direction as a three-vector `[n m p]`.

Example 9.9

Let the solid be the unit cube, with corner at the origin, $R = \{0 \leq x, y, z \leq 1\}$, and let $f(x, y, z) = (y + z) \sin(\pi(x^2 + y^2))$. First we do the integral using the symbolic integrator three times. We then do the integral using the mfile `simp3`.

```
>> syms x y z
>> ff = (y+z)*sin(pi*(x^2 +y^2))
>> int(int(int(ff, x,0,1), y,0,1),0,1)
ans =
1/2*FresnelC(2^(1/2))*(FresnelS(2^(1/2))*pi+2^(1/2))/pi
>> double(ans)
ans =
0.30784938470884

>> f =
inline('(y+z).*sin(pi*(x.^2 +y.^2))', 'x', 'y', 'z')
>> corners = [0 1 0 1 0 1]
>> simp3(f, corners)
enter the number of subdivisions [n m p] [20 20 20]
ans =
0.30786314294381

>> simp3(f, corners)
enter the number of subdivisions [n m p] [40 40 40]
ans =
0.30785023200358
```

The result for the symbolic calculation is given in terms of the following Fresnel functions:

$$\text{Fresnel}C(x) = \sqrt{\frac{2}{\pi}} \int_0^{\sqrt{x}} \cos(t^2)\, dt$$

$$\text{Fresnel}S(x) = \sqrt{\frac{2}{\pi}} \int_0^{\sqrt{x}} \sin(t^2)\, dt.$$

The three answers differ by an amount on the order of 10^{-6}.

Integration over nonrectangular regions is accomplished by using a transformation T to pull back the region and function to a rectangular set, as in the two-dimensional setting. Let the transformation

$$(x, y, z) = T(u, v, w) = (\alpha(u, v, w), \beta(u, v, w), \gamma(u, v, w))$$

map the solid rectangle R onto the set G in x, y, z space. The Jacobian matrix of T is

$$J(u, v, w) = \begin{bmatrix} \alpha_u & \alpha_v & \alpha_w \\ \beta_u & \beta_v & \beta_w \\ \gamma_u & \gamma_v & \gamma_w \end{bmatrix}.$$

We let $|J(u, v, w)|$ stand for the absolute value of $\det(J(u, v, w))$. The formula for change of variable is

$$\int \int \int_G f(x, y, z) \, dV(x, y, z)$$
$$= \int \int \int_R f(\alpha(u, v, w), \beta(u, v, w), \gamma(u, v, w)) |J(u, v, w)| \, dV(u, v, w).$$
$$(9.13)$$

Spherical coordinates

Integration in regions described in spherical coordinates uses Eq. (9.13). Here the change of variable is

$$(x, y, z) = T(r, \theta, \phi),$$

where

$$\begin{aligned} x &= r \cos\theta \sin\phi \\ y &= r \sin\theta \sin\phi \\ z &= r \cos\phi. \end{aligned}$$

The radial distance is $r = \sqrt{x^2 + y^2 + z^2}$, the angle in the xy plane is θ, and the angle from the north pole is ϕ. The Jacobian matrix for this T is

$$J(r, \theta, \phi) = \begin{bmatrix} \cos\theta \sin\phi & -r \sin\theta \sin\phi & r \cos\theta \cos\phi \\ \sin\theta \sin\phi & r \cos\theta \sin\phi & r \sin\theta \cos\phi \\ \cos\phi & 0 & -r \sin\phi \end{bmatrix}.$$

The determinant of the Jacobian matrix is $|J(r, \theta, \phi)| = r^2 \sin \phi$, so Eq. (9.13) becomes

$$\int \int \int_G f(x, y, z)dV(x, y, z) = \int \int \int_R \tilde{f}(r, \theta, \phi)r^2 \sin \phi \, dr \, d\theta \, d\phi, \quad (9.14)$$

where the pullback

$$\tilde{f}(r, \theta, \phi) = f(r \cos \theta \sin \phi, \; r \sin \theta \sin \phi, \; r \cos \phi).$$

Example 9.10

Let G be the region between the spheres of radius $r = 2$ and $r = 4$, centered at the origin. We suppose the region is filled with a material of variable density $\rho(x, y, z) = 1 + \cos(x)$. What is the mass of the region, and what is the average density? The mass is given by

$$\int \int \int_G \rho(x, y, z) \, dV(x, y, z).$$

The region between the spheres is described by $2 \leq r \leq 4$, $0 \leq \theta \leq 2\pi$, and $0 \leq \phi \leq \pi$. Using the change-of-variable formula (9.14), we have

$$\int \int \int_G \rho(x, y, z) \, dV(x, y, z)$$

$$= \int_0^\pi \int_0^{2\pi} \int_2^4 [1 + \cos(r \cos \theta \sin \phi)]r^2 \sin \phi \, dr \, d\theta \, d\phi.$$

We estimate this integral using the mfile `simp3`:

```
function out = rrho(r,theta,phi)
    x = r.*cos(theta).*sin(phi);
    out = (1+cos(x)).*r.^2.*sin(phi);
```

. . . .

```
>> corners = [2 4 0 2*pi 0 pi]

>> simp3('rrho', corners)
enter the number of subdivisions [n m p] [20 60 30]
ans =
2.360351669395974e+02
```

```
>> simp3('rrho', corners)
enter the number of subdivisions [m n p]  [40 120 60]
ans =
2.360324035495266e+02
```

The difference in the answers is on the order of 3×10^{-3}. We conclude that the absolute error in the first answer is on this order. The *relative* error is on the order of $3 \times 10^{-3}/200 = 1.5 \times 10^{-5}$. The average density is found by dividing the mass by the volume, $(4/3)\pi(4^3 - 2^3)$.

Exercises

1. a) Let R be the rectangle $[0, 2] \times [0, 2]$ and $f(x, y) = \sin(x + y)$. Use the mfile `riemann` with $m = n = 4$, $m = n = 8$, $m = n = 16$, and $m = n = 32$, and each time use the option to show the graph of the step function approximation.

b) Calculate the exact result by hand. What is the difference between the Riemann sum with $m = n = 32$ and the exact result?

2. a) Let G be the set $\{(x, y) : y > x^2/2\} \cap R$, where R is the square of Exercise 1. We shall try the direct Riemann sum approach to estimate the integral

$$\int \int_G \sin(x + y) \, dA(x, y).$$

The restricted function $\tilde{f}$ is

$$\tilde{f}(x, y) = \begin{cases} \sin(x + y) & \text{for } y > x^2/2 \\ 0 & \text{for } y < x^2/2 \end{cases}$$

The function mfile for $\tilde{f}$ is

```
function z = ftil(x,y)
        z = (y >.5*x.^2).* sin(x+y);
```

Use the mfile `riemann` on $\tilde{f}$ with $m = n = 50$.

b) Now evaluate the integral as an iterated integral

$$\int \int_G \sin(x + y) \, dA(x, y) = \int_0^2 \int_{x^2/2}^2 \sin(x + y) \, dy \, dx$$

using the symbolic integrator. The answer is given in terms of the Fresnel functions. Evaluate the answer in double precision. What is the difference between this answer and the answer in part a)? The evaluation in terms of Fresnel integrals is more accurate and may be taken as the "exact answer."

3. Recall that the trapezoid rule in one dimension is

$$T_n(f) = \left(\frac{b-a}{2n}\right)[f(x_1) + 2f(x_2) + \cdots + 2f(x_n) + f(x_{n+1})]$$

and

$$\int_a^b f(x)\,dx = T_n(f) + E_n(f).$$

Letting $h = (b-a)/n$, the error may be expressed as

$$E_n(f) = -\frac{1}{12}(b-a)h^2 f''(\xi)$$

for some point $\xi \in [a, b]$.

a) What is the two-dimensional version of the trapezoid rule on a rectangle $R = [a, b] \times [c, d]$? How can this rule be interpreted on each subrectangle?

b) Let $T_{n,m}$ be the two-dimensional trapezpoid rule, where n is the number of subdivisions in the x direction and m is the number of subdivisions in the y direction. Show that for a function of the form $f(x, y) = g(x)h(y)$,

$$T_{n,m}(f) = T_n(g)T_m(h).$$

c) For a function of the form $f(x, y) = g(x)h(y)$, we know that

$$\int\int_R f(x, y)\,dA(x, y) = \int_a^b f(x)\,dx \int_c^d g(y)\,dy.$$

Show that the error in the two-dimensional trapezoid method is

$$E = E_n(g)T_m(h) + T_n(g)E_m(h) + E_n(g)E_m(h).$$

How should the error decrease when n and m are doubled?

d) Write a MATLAB script that implements the two-dimensional trapezoid rule in the manner of the two-dimensional Simpson's rule.

e) Try out your script on $f(x, y) = x + y^2$ on the unit square $R = [0, 1] \times [0, 1]$. Calculate the exact value of the integral by hand. Then calculate the actual error for various choices of n and m. How does the error decrease when n and m are doubled?

4. Let R be the rectangle $[0, 2] \times [1, 4]$.

a) Let $f(x, y) = x \cos(x^2 + y)$. Calculate the integral $\int \int_R f \, dA$ by hand to get the exact value.

b) Use the code `simp2` to estimate the integral with $[n, m] = [40, 60]$. Call the result I_1. Then estimate it again with $[n, m] = [80, 120]$ and call the result I_2. How well does the difference $|I_1 - I_2|$ estimate the error $I_1 - \int \int_R f \, dA$?

5. Let R be the rectangle $[0, 5] \times [0, 5]$ and let $f(x, y) = \exp(x + \sin y)$. Use `simp2` to estimate the integral $\int \int_R f \, dA$, first with $[n, m] = [50, 50]$ and then with $[n, m] = [100, 100]$. Estimate the error in the first result. Finally use `dblquad`. Compare the results.

6. A thin metal plate of constant density occupies the set G in the xy plane, where $G = \{(x, y) : 2x^3 - 1 \le y \le 2(x - .5)^2 + .5, \ 0 \le x \le 1\}$. The center of mass of the plate has coordinates $\bar{x}, \bar{y}$ where

$$\bar{x} = \frac{\int \int_G x \, dA(x, y)}{\int \int_G dA(x, y)}, \qquad \bar{y} = \frac{\int \int_G y \, dA(x, y)}{\int \int_G dA(x, y)}.$$

a) Graph the plate in the xy plane.

b) Use the symbolic integrator to find $\bar{x}, \bar{y}$.

7. Let the transformation T be given by $x = f(u, v) = \exp(u) \cos v$ and $y = g(u, v) = \exp(u) \sin v$. Let R be the rectangle $\{0 \le u \le 2, \ 0 \le v \le \pi\}$.

a) Let G be the image of R under this transformation. Sketch the set G in the xy plane. What is the area of G?

b) Now use the mfile `trf.m` on this problem, first with $m = n = 2$, then with $m = n = 4$, and then with $m = n = 8$. Compare the area of the approximating set of parallelograms with the exact answer of part a). How large must you choose m, n to get within .15 of the exact answer?

8. Let the transformation T be given by $x = f(u, v) = u + v/2$ and $y = g(u, v) = v \exp(u/2)$. Let R be the rectangle $\{0 \le u \le 4, \ -1 \le v \le 1\}$.

a) Let G be the image of R under the transformation T. Sketch the set G in the xy plane. Calculate its area exactly using the change-of-variable formula (9.11).

b) Now use the mfile `trf.m` on this problem, first with $m = n = 2$, then $m = n = 4$, and finally $m = n = 8$. Compare the area of the approximating set of parallelograms each time with the exact area found in part a). How large must you choose m, n to get within .01 of the exact answer?

9. Find the volume of the portion of the ball $x^2 + y^2 + z^2 \le 4$ that lies above the surface $z = \exp(x)$.

a) Set up as a double integral over the set $G = \{x^2 + \exp(2x) + y^2 \leq 4\}$.

b) Find out what the boundary of the set G looks like by graphing $y = g(x) = \pm\sqrt{4 - x^2 - \exp(2x)}$ on the interval $-2 \leq x \leq 1$.

c) We can write the integral over G as an iterated integral for an appropriate f as

$$\int_a^b \int_{-g(x)}^{g(x)} f(x, y)\, dy\, dx = 2 \int_a^b \int_0^{g(x)} f(x, y)\, dy\, dx.$$

a and b are zeros of g. Use `fzero` and the function $g(x)$ to find a and b.

d) Make the change of variable $x = u$, $y = v\sqrt{4 - u^2 - \exp(2u)}$ to bring the integral into the form

$$2 \int_a^b \int_0^1 q(u, v)\, dv\, du.$$

e) Use `simp2` first with $m = n = 50$ and then with $m = n = 100$. Estimate the error in your first result.

10. Let G be the ellipsoid $\{x^2 + y^2/4 + z^2/9 \leq 1\}$. Suppose that a material with density $\rho(x, y, z) = \exp(-x^2 - y^2 - z^2)$ occupies this region.

a) Graph the body by graphing the functions $z = \pm 3\sqrt{1 - x^2 - y^2/4}$ over the set $\{x^2 + y^2/4 \leq 1\}$.

b) Find the mass of the body. First make the change of variable $x = u$, $y = 2v$, $w = 3z$ to transform G into the unit ball. Then use spherical coordinates and the mfile `simp3`.

11. This problem is a slight variation on a standard problem in multivariable calculus. However, in this variation, the integral cannot be done analytically.

Find the volume of the region that is the intersection of the solid cylinder $\{(x - 1)^2 + y^2 \leq 1/4\}$ with the ball $\{x^2 + y^2 + z^2 \leq 4\}$. Make a change of variable $\tilde{x} = x - 1$ to move the cylinder to have as its axis of symmetry the z axis. Use polar coordinates and then use `simp2` to estimate the integral with an error on the order of 10^{-6}.

* **12.** Consider a solid cylinder H of radius $b < 2$, aligned along the x axis, $H = \{(x, y, z) : y^2 + z^2 \leq b^2\}$. Suppose this cylinder consists of a material with density $\rho(x, y, z) = 1/(1 + x^2)$.

a) Find the mass of the part of the cylinder in the interval $|x| \leq 4$. This can be done by hand easily.

b) Now cut out a cylinder of radius $a < b$, with its axis of symmetry the z axis. Let this set be denoted $F = H \cap \{x^2 + y^2 \leq a^2\}$. Graph the two cylinders for the choice of parameters $a = 1$, $b = 2$.

c) To find the mass of F, we must compute the integral $\int \int \int_F \rho(x)dxdydz$. Do the z integration by hand, reducing the problem to that of evaluating a double integral over the disk $\{x^2 + y^2 \le b^2\}$. Introduce polar coordinates, and then use simp2 to estimate the value of this integral for the case $a = 1$, $b = 2$.

d) Now, to make the problem more interesting, suppose the smaller cylinder has the line $z = x$, $y = 0$ as its axis of symmetry. Thus $F = H \cap \{(x-z)^2 + 2y^2 = 2a^2\}$. We want to compute the mass of F. To reduce this problem to that of part b), set $\tilde{x} = x - z$ and leave y and z unchanged. This change of variable maps the cylinder $\{y^2 + z^2 \le b^2\}$ onto itself, but changes the density into $\tilde{\rho}(\tilde{x}, y, z) = 1/(1 + (\tilde{x} + z)^2)$. It also transforms the smaller-diameter cylinder into a cylinder with its axis of symmetry the z axis but with elliptical cross section $\tilde{x}^2 + 2y^2 = 2a^2$. Make a second change of variable to map the smaller cylinder into a cylinder with a circular cross section. Estimate the value of the integral using dblquad.

★   13. Let a circular helix be given by

$$\mathbf{p}(t) = (r \cos t, \; r \sin t, \; at), \quad 0 \le t \le 8\pi.$$

A tube of circular cross section of radius $b < r$ follows the helix with the circular cross sections perpendicular to the tangent vector of the helix. Think of a coiled tube.

a) Make the following change of variable,

$$(u, v, t) \rightarrow (x, y, z) = \mathbf{p}(t) + u\mathbf{N}(t) + v\mathbf{B}(t),$$

where $\mathbf{N}(t) = (-\cos t, -\sin t, 0)$ is the principal normal at $\mathbf{p}(t)$ and where $\mathbf{B}(t) = (a \sin t, -a \cos t, r)/\sqrt{r^2 + a^2}$ is the binormal at $\mathbf{p}(t)$. This change of variable pulls back the coiled tube to the cylinder $\{u^2 + v^2 \le b^2, \; 0 \le t \le 8\pi\}$ (cf. Exercise 10 of Chapter 6). Show that for this change of variable, $|J(u, v, t)| = (a^2 + r(r + u))/\sqrt{r^2 + a^2}$.

b) Show that the volume of the coiled tube is $8\pi^2 b^2 \sqrt{r^2 + a^2}$. This calculation can be done easily by hand.

c) Now suppose that the coiled tube is filled with a liquid containing metallic particles. A magnetic field is applied that affects the density of metal particles. It becomes $\delta(x, y, z) = (x + 10)^{3/2}$. Neglecting the mass of the fluid in the tube, calculate the mass of metal particles in the tube. Make the change of variable of part a), and write the integral that must be calculated. Set $r = 4, a = 1$, and $b = .5$ and estimate the integral using simp3.

★★   14. Find the volume of your dog.

10

Scalar Integrals Over Curves and Surfaces

Prepared mfiles used in this chapter

simp2   tsurf   gdome

10.1   Scalar integrals along curves

The integral of a scalar-valued function along a curve C arises in many contexts. These integrals have the form

$$\int_C f\, ds,$$

where s is arc length and f is some function defined on the curve. The function f may be a function of the arc length, or it may be a function of the coordinates of a point on C.

Example 10.1

Suppose that the density of a curved wire depends on the arc length from one end of the wire. Then the mass of the wire is given by the integral

$$m = \int_C \rho(s)\, ds = \int_0^L \rho(s)\, ds,$$

where $\rho(s)$ is the linear density and L is the length of the curve.

197

To find the coordinates of the center of mass of this piece of wire, we must evaluate the integrals

$$\bar{x} = \frac{1}{m} \int_C x\rho(s)\, ds$$

$$\bar{y} = \frac{1}{m} \int_C y\rho(s)\, ds$$

$$\bar{z} = \frac{1}{m} \int_C z\rho(s)\, ds.$$

Let C be the semicircle of radius 2 in the upper half-plane, and suppose that the density $\rho(s) = \sqrt{1+s^2}$, where the arc length is measured from the point $(2, 0)$. The arc length of the semicircle is $L = 2\pi$ so that the mass is given by

$$m = \int_0^{2\pi} \sqrt{1+s^2}\, ds$$

This integral can be done by hand or symbolically.

```
>> syms s
>> density = sqrt(1+s^2)
>> int(density, 0, 2*pi)
1/2*pi*(1+pi^2)^(1/2)-1/2*log(-pi+(1+pi^2)^(1/2))
>> mass = double(ans)
mass =
  21.2563
```

Now, to calculate the coordinates of the center of mass we parameterize C by

$$\mathbf{r}(t) = [2\cos(t),\ 2\sin(t),\ 0], \qquad 0 \le t \le \pi.$$
$$ds = \|\mathbf{r}'(t)\|dt = 2dt.$$

Then $s(t) = 2t$ and $\rho(s(t)) = \sqrt{1+4t^2}$. Since $x = 2\cos(t)$ on the curve, we have

$$\bar{x} = (1/m) \int_0^\pi 2\cos(t)\sqrt{1+4t^2}\, 2dt.$$

This integral cannot be done symbolically, so we use `quadl`:

```
>> f = inline('4*sqrt(1+4*t.^2).*cos(2*t)')
>> xbar = (1/mass)*quadl(f,0, pi)
xbar =
  0.1267
```

10.2   Scalar integrals on surfaces

If the surface S is given as the graph of a continuously differentiable function $z = f(x, y)$ over a set G, the formula for the surface area is

$$\text{Area}(S) = \int\int_G \sqrt{1 + f_x^2 + f_y^2} \, dA(x, y). \tag{10.1}$$

If $g(x, y, z)$ is a function defined in a region in x, y, z space containing the piece of surface over G, the integral of g over S is defined to be

$$\int\int_G g(x, y, f(x, y))\sqrt{1 + f_x^2 + f_y^2} \, dA(x, y). \tag{10.2}$$

We often write the element of surface area as

$$dS = \sqrt{1 + f_x^2 + f_y^2} \, dA(x, y),$$

so we may write the scalar surface integral as

$$\int\int_S g(x, y, z) \, dS.$$

Example 10.2

Let the surface S be given by $z = f(x, y) = \sin(x + y^2)$ over the square $R = [0, 2] \times [0, 2]$. The area of this piece of surface is

$$\int\int_R \sqrt{1 + \cos^2(x + y^2)(1 + 4y^2)} \, dA(x, y).$$

Frequently this kind of integral cannot be done symbolically, so we estimate the integral using the two-dimensional version of Simpson's rule:

```
>> h =
inline('sqrt(1 + cos(x+y.^2).^2.*(1 + 4*y.^2))', 'x', 'y')
>> simp2(h, [0 2 0 2])
enter the number of subdivisions [n m]   [20 20]
ans =
7.2589
```

Suppose the surface S is a sheet of metal with a variable density $\rho(x, y, z)$. The mass of the sheet of metal is given by the scalar surface integral

$$m = \int\int_S \rho(x, y, z)\, dS = \int\int_R \rho(x, y, f(x, y))\sqrt{1 + f_x^2 + f_y^2}\, dA(x, y).$$

The coordinates of the center of mass are given by the integrals

$$\bar{x} = (1/m)\int\int_S x\rho(x, y, z))\, dS$$

$$\bar{y} = (1/m)\int\int_S y\rho(x, y, z)\, dS$$

$$\bar{z} = (1/m)\int\int_S z\rho(x, y, z))\, dS.$$

Example 10.3

We continue with the surface of Example 10.2 over the set $G = R = [0, 2] \times [0, 2]$. Suppose the density of the material is given by $\rho(x, y, z) = x + y + z$. Then the mass of the sheet of metal is given by the integral

$$m = \int\int_R (x + y + \sin(x + y^2))\sqrt{1 + \cos^2(x + y^2)(1 + 4y^2)}\, dA(x, y).$$

The z coordinate of the center of mass is given by

$$(1/m)\int\int_R (x + y + \sin(x + y^2))\sin(x + y^2)\sqrt{1 + \cos^2(x + y^2)(1 + 4y^2)}\, dA(x, y).$$

Just for practice, we estimate these integrals using the mfile dblquad. Since it is not possible to combine inline functions, we write an mfile for the integrands:

```
function integrand = h(x,y)
    z = sin(x+y.^2);
    density = x+y+z;
    area_element = sqrt(1 + cos(x+y.^2).^2.*(1+4*y.^2));
    integrand = density.*area_element;
```
. . . .
```
function integrand = h1(x,y)
    z = sin(x+y.^2);
    integrand = z.*h(x,y);
```
. . . .

```
>> mass = dblquad('h', 0, 2, 0, 2)
mass = 17.0008

>> zbar = (1/mass)*dblquad('h1', 0, 2, 0, 2)
zbar =   .2285
```

10.3   Integrals over surfaces given parametrically

Recall that a surface is represented parametrically when it is the image set in xyz space of a set R (usually a rectangle) in uv space under the three functions

$$(u, v) \to (x(u, v), \ y(u, v), \ z(u, v)).$$

How do we calculate the area of such a surface, and how do we calculate integrals of scalar functions over them?

The element of surface area for a surface S represented parametrically is defined to be

$$dS = \|\mathbf{n}(u, v)\| \, dA(u, v)$$

$$= \sqrt{(y_u z_v - y_v z_u)^2 + (x_v z_u - x_u z_v)^2 + (x_u y_v - x_v y_u)^2} \, dA(u, v).$$

The area of the surface is therefore

$$\text{Area}(S) = \int \int_R \|\mathbf{n}(u, v)\| \, dA(u, v) \tag{10.3}$$

and the integral of a function $f(x, y, z)$, such as a density, is

$$\int \int_S f(x, y, z) \, dS = \int \int_R f(x(u, v), y(u, v), z(u, v)) \|\mathbf{n}(u, v)\| \, dA(u, v). \tag{10.4}$$

In the case of a surface of revolution, $\mathbf{n}$ is given in Eq. (6.4), and the element of surface area is

$$dS = \|\mathbf{n}(u, v)\| \, dA(u, v) = |x_0(v)| \sqrt{[x_0'(v)]^2 + [z_0'(v)]^2} \, du dv.$$

Hence the area of a surface of revolution reduces to a single integral in v,

$$\text{Area}(S) = 2\pi \int_a^b |x_0(v)| \sqrt{[x_0'(v)]^2 + [z_0'(v)]^2} \, dv. \tag{10.5}$$

Example 10.4

We shall find the surface area of a banana. First we must find a parameterization of what we think a banana should look like. Begin with an ellipsoid, which is a surface of revolution.

$$x(u, v) = \cos u \cos v$$
$$y(u, v) = \sin u \cos v$$
$$z(u, v) = 3 \sin v, \qquad 0 \le u \le 2\pi, \qquad -\pi/2 \le v \le \pi/2.$$

Then bend it over in the x direction by adding on a perturbation to the x coordinate function,

$$x(u, v) = \cos u \cos v + 2 \sin^2 v.$$

The ellipsoid and the banana are shown in Figure 10.1, left and right. The normal vector is

$$\mathbf{n}(u, v) = [3 \cos u \cos^2 v, \; 3 \sin u \cos^2 v, \; \cos v \sin v - 4 \cos u \sin v \cos^2 v],$$

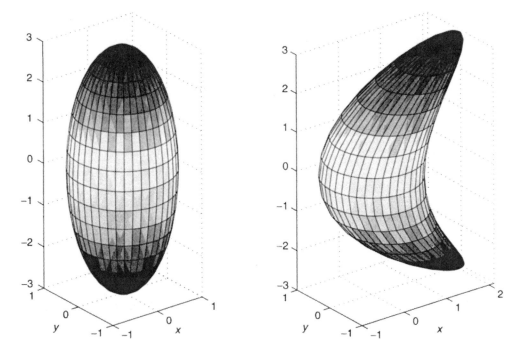

Figure 10.1 Ellipsoid, on the left, and banana, on the right.

and the area is given by the integral

$$\int_0^{2\pi} \int_{-\pi/2}^{\pi/2} \|\mathbf{n}(u, v)\| \, dA(u, v),$$

where

$$\|\mathbf{n}(u, v)\|^2 = 9\cos^4 v + (\cos v \sin v - 4\cos u \sin v \cos^2 v)^2.$$

The graphing and the computation are done as follows: The expression for $\|\mathbf{n}\|$ is rather complicated, so we write it in an mfile:

```
function out = da(u,v)
n1 = 3*cos(u).*cos(v).^2;
n2 = 3*sin(u).*cos(v).^2;
n3 = cos(v).*sin(v) - 4*cos(u).*sin(v).*cos(v).^2;
out = sqrt(n1.^2 + n2.^2 +n3.^2);
```

```
% First we graph the banana.
>> u = linspace(0,2*pi,41)
>> v = linspace(-.5*pi,.5*pi, 41)
>> [U,V] = meshgrid(u,v);
>> Y = cos(U).*cos(V) + 2* sin(V).^2;
>> Y = sin(U).*cos(V);
>> Z = 3*sin(V);
>> surf(X,Y,Z);
% Now we compute the surface area.
>> simp2('da', [0 2*pi .5*pi .5*pi])
enter the number of subdivisions [n,m], [50 50]
ans =
   33.37013464057785

simp2('da', [0 2*pi -.5*pi .5*pi])
enter the number of subdivisions [n,m] [100 100]
ans =
   33.37016362462368
```

Comparing the first and second Simpson estimates, we see that the first estimate probably has an error on the order of 3×10^{-5}.

10.4 Surfaces composed of triangles

It is often interesting, and more convenient, to construct surfaces from triangular patches. This is done in many contexts, including computer graphics and mechanical engineering.

Computer graphics

MATLAB uses a triangular patch representation of the surface in the surf command. After a mesh (x_j, y_i) is put on the rectangle R with the meshgrid command, we create a matrix of z values, $z_{i,j} = f(x_j, y_i)$, with the command Z = f(X,Y). It appears that the surface is constructed from small parallelograms with vertices at the points in space $(x_j, y_i, f(x_j, y_i))$. However, we know that in general a plane cannot be fit through four distinct points in space. If we look closely at a test case, we can see that what appears to be a parallelogram is really two triangles. In Figure 10.2, we have used the surf command to produce a surface through the four points in space,

$$P_1 = (0, 0, 0), \qquad P_2 = (1, 0, 0), \qquad P_3 = (1, 1, 1), \qquad P_4 = (0, 1, 0).$$

These four points do not lie in a plane. The surface in Figure 10.2 consists of one triangle with vertices at P_1, P_2, P_3 and a second triangle with vertices at P_1, P_3, P_4.

Geodesic domes

The roofs of many structures are surfaces constructed of triangles and other polygon shapes. For ease of construction it is desirable that the triangles have only one or two

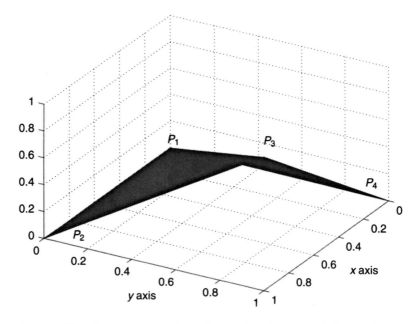

Figure 10.2 Surface consisting of two triangles through four points.

shapes, preferably equilateral or isosceles. An important class of these triangulated surfaces is the geodesic dome introduced by R. Buckminster Fuller. A geodesic dome is very strong and light, and no interior columns are needed to support it.

The simplest geodesic structure is the icosahedron, consisting of a 20-sided polyhedron with all faces equal to the same equilateral triangle. Of course, the dome would be the upper half of the icosahedron. The icosahedron can be refined by dividing each of the triangular faces into 6 congruent right triangles, thereby yielding a polyhedron with 120 identical faces.

One of Fuller's designs started with a polyhedron consisting of 32 faces, of which 12 are pentagons, interspersed with 20 faces that are hexagons. This polyhedron is known as a Bucky ball. The subdivisions of a soccer ball also have this form. Each of the hexagons may be subdivided into 6 triangles, and each of the pentagons into 5 triangles, for a total of 180 triangles. They are all isosceles triangles, even though the triangles of the hexagons may appear to be equilateral.

To see this object, use the mfile gdome.m. Enter gdome. The initial polyhedron, which looks like a soccer ball, is displayed. Then enter Return. The hexagons are divided into triangles. Enter Return a third time to see the pentagons divided into triangles.

Surface area by triangles

We can also approximate a smooth surface with triangular patches. The area of the smooth surface can be approximated easily by adding up the areas of the triangular patches. Of course, this will not yield the exact area of the surface as given by the surface area integral (Eq. 10.1). It is only an approximation. Furthermore, if we choose the triangular patches differently, we will probably get a different answer. However, computing the area of this approximation to the surface is easier than using the integral formula, because we do not need to know the values of the derivatives of the function.

We illustrate this kind of approximation in a very simple example. Let $f(x, y)$ be defined on the rectangle $R = \{a \leq x \leq b, \ c \leq y \leq d\}$. We construct a triangular patch approximation to the graph of f using only the values at the four corners. We set

$$
\begin{aligned}
z_1 &= f(a, c) \\
z_2 &= f(b, c) \\
z_3 &= f(b, d) \\
z_4 &= f(a, d)
\end{aligned}
$$

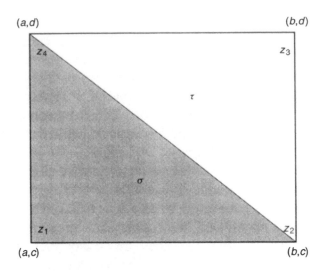

Figure 10.3 Rectangle divided into two triangles.

Then we divide the base rectangle R into two triangles, as in Figure 10.3. Over each of these triangles, there is a triangular surface patch. Let σ be the triangular patch with vertices

$$(a, c, z_1), \qquad (b, c, z_2), \qquad (a, d, z_4),$$

and let τ be the triangular patch with vertices

$$(b, c, z_2), \qquad (b, d, z_3) \qquad (a, d, z_4).$$

The edge vectors of σ, attached to the vertex (a, c, z_1), are

$$\mathbf{u} = [b - a, \ 0, \ z_2 - z_1], \qquad \mathbf{v} = [0, \ d - c, \ z_4 - z_1].$$

Hence the area of σ is

$$(1/2)\|\mathbf{u} \times \mathbf{v}\| = (1/2)\sqrt{(b - a)^2(d - c)^2 + (b - a)^2(z_4 - z_1)^2 + (d - c)^2(z_2 - z_1)^2}.$$

In similar fashion, the area of triangle τ is found to be

$$(1/2)\sqrt{(b - a)^2(d - c)^2 + (b - a)^2(z_2 - z_3)^2 + (d - c)^2(z_4 - z_3)^2}.$$

The area of the surface consisting of these two patches is just the sum of the areas of σ and τ.

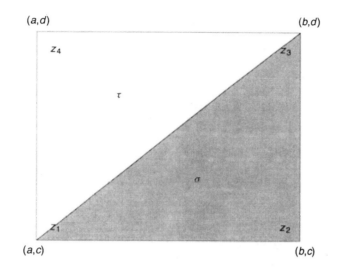

Figure 10.4 Alternate way of dividing a rectangle into a triangle.

It is also possible to divide the base rectangle into two triangles, as in Figure 10.4. With σ and τ as the triangular patches over these base triangles, the formula is

$$\text{area}(\sigma) = (1/2)\sqrt{(b-a)^2(d-c)^2 + (d-c)^2(z_2-z_1)^2 + (b-a)^2(z_3-z_2)^2}$$

and

$$\text{area}(\tau) = (1/2)\sqrt{(b-a)^2(d-c)^2 + (d-c)^2(z_4-z_3)^2 + (b-a)^2(z_4-z_1)^2}.$$

Example

Let $f(x, y) = xy$ over the rectangle $R = \{0 \le x, y \le 1\}$. Then $z_1 = z_2 = z_4 = 0$ and $z_3 = 1$. The triangulation depicted in Figure 10.3 yields a surface area approximation of $(1 + \sqrt{3})/2 \approx 1.3660$. The other triangulation yields a surface area approximation of $\sqrt{2} \approx 1.414$. The true surface area is

$$\int_0^1 \int_0^1 \sqrt{1 + f_x^2 + f_y^2}\, dx\, dy = \int_0^1 \int_0^1 \sqrt{1 + x^2 + y^2}\, dx\, dy = 1.2808.$$

More generally, let S be given as the graph of a function $z = f(x, y)$ over a rectangle $R = [a, b] \times [c, d]$. We shall calculate the area of a triangular patch approximation to S as follows. Introduce a mesh on R with meshpoints (x_j, y_i),

$$a = x_1 < \cdots < x_{n+1} = b, \qquad c = y_1 < \cdots < y_{m+1} = d$$

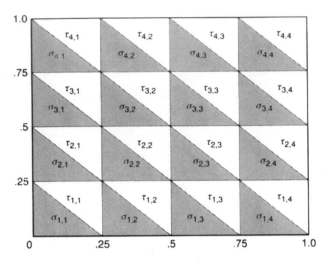

Figure 10.5 Rectangle divided into 32 triangles.

and values $z_{i,j} = f(x_j, y_i)$ at the meshpoints. Let $R_{i,j}$ be the subrectangle $x_j \leq x \leq x_{j+1}$, $y_i \leq y \leq y_{i+1}$. Dividing each $R_{i,j}$ into triangles $\sigma_{i,j}$ and $\tau_{i,j}$ (see Figure 10.5) we see that the area of the surface consisting of triangular patches is

$$A = \sum_{i=1}^{m} \sum_{j=1}^{n} [\text{area}(\sigma_{i,j}) + \text{area}(\tau_{i,j})] \tag{10.6}$$

where

$$\text{area}(\sigma_{i,j}) = (1/2)\sqrt{\Delta x^2 \Delta y^2 + \Delta x^2 (z_{i+1,j} - z_{i,j})^2 + \Delta y^2 (z_{i,j+1} - z_{i,j})^2}$$

$$\text{area}(\tau_{i,j}) = (1/2)\sqrt{\Delta x^2 \Delta y^2 + \Delta x^2 (z_{i,j+1} - z_{i+1,j+1})^2 + \Delta y^2 (z_{i+1,j} - z_{i+1,j+1})^2}.$$

mfile tsurf

The mfile `tsurf` produces a triangular patch approximation to a surface and computes the sum of the areas of the triangles. If X, Y is a mesh on the rectangle R and Z = f(X,Y), the call is `tsurf(X,Y,Z)`. You will see the triangles displayed in two colors, blue and cyan.

Example 10.5

Let $f(x, y) = x^2 + y^2$ on the square $[0, 2] \times [0, 2]$. We make a coarse mesh with only two subdivisions in each direction. We also graph f with the `surf` command

and a finer mesh to see how the triangles fit the surface. The results are displayed in Figure 10.6. We also estimate the area integral (Eq. 10.1) with `simp2`:

```
>> f = inline('x.^2 + y.^2', 'x', 'y')
% We compute the area of the triangular patchs.
>> [X,Y] = meshgrid(0:2);
>>   Z = f(X,Y);
>>   tsurf(X,Y,Z)
ans =
12.724

>> hold on
% We graph the surface with a finer mesh.
>> [XX,YY] = meshgrid(0:.1:2);
>> ZZ = f(XX,YY);
>> surf(XX,YY,ZZ); colormap(gray); shading interp;


% We compute the surface area integral numerically.
>> g = inline('sqrt(1 + 4*(x.^2 ı y.^2))', 'x', 'y')
>> corners = |0 2 0 2|
>> simp2(g, corners)
enter the number of subdivisions [n m]   [40 40]
ans =
13.004
```

Figure 10.6 Surface $z = x^2 + y^2$, on the left, and triangular approximation, on the right.

If we continue to increase the number of triangles and do not allow them to become too "skinny," the triangular patch approximation will converge to the surface area, for any triangulation of the rectangle.

The minimal-surface problem

A basic problem in geometry, which has consequences in engineering and architecture, is that of finding the surface of least area that spans a given wire frame. Such surfaces are easy to produce physically by dipping a wire frame in a soapy solution. The resulting soap films are minimal surfaces. There are very few examples of minimal surfaces that are expressible in mathematically closed form. One of these is the *catenoid*. It is the surface of revolution that is formed by revolving the curve $y = c \cosh(x/c)$ about the x axis.

In a particular kind of minimal-surface problem, we specify the height of the surface around the boundary of a set G. This is called *Plateau's problem* and it can be stated in precise mathematical terms as follows.

Given a set G in x, y space, and a function f defined on the boundary of G, let $\mathcal{A}$ be the set of all differentiable functions $u(x, y)$ on G such that $u = f$ on the boundary of G. Then we seek that function $u \in \mathcal{A}$ that minimizes the surface area:

$$\int \int_G \sqrt{1 + u_x^2 + u_y^2} \, dA.$$

If the minimizing function u has continuous second order partial derivatives, it can be shown that u must satisfy the nonlinear partial differential equation

$$\frac{\partial}{\partial x} \left(\frac{u_x}{\sqrt{1 + u_x^2 + u_y^2}} \right) + \frac{\partial}{\partial y} \left(\frac{u_y}{\sqrt{1 + u_x^2 + u_y^2}} \right) = 0. \qquad (10.7)$$

Equation (10.7) is called the minimal-surface equation. It is very hard to construct and analyze solutions of this equation. We shall instead use an engineering approach to Plateau's problem over a rectangle R in which we attempt to approximate the minimal surface with a surface consisting of triangles. In this context, the triangle surface patches are called *finite elements*.

Let R be a rectangle in x, y space, $a \leq x \leq b$, $c \leq y \leq d$. We could specify four separate functions of one variable, one for each side of the rectangle. Instead we shall consider one function, $f(x, y)$, defined on R, and then use its restriction to the four sides of R as the boundary values for the surface. We shall construct a surface

S, consisting of triangles, that agrees with f on the sides of R. We shall then adjust the triangles in the interior of R to make the area of S as small as possible.

We introduce meshpoints (x_j, y_i) in the rectangle, as before, and calculate the area of the surface consisting of triangles using Eq. (10.6). The values $z_{i,j} = f(x_j, y_i)$ are given for (x_j, y_i) on the boundary of R. However, for the interior points, indicated in Figure 10.7, we can choose the values. This means that the area formula (10.6) defines a function of the $(n-1)(m-1)$ variables $z_{i,j}$, with $i = 2, \ldots, m$, $j = 2, \ldots, n$,

$$A = A(z_{2,2}, z_{2,3}, \ldots, z_{2,n}, z_{3,2}, \ldots, z_{3,n}, \ldots, z_{m,2}, \ldots, z_{m,n}).$$

We illustrate with an example where the numbers are not too large.

Example 10.6

Let R be the square $[1, 4] \times [1, 4]$. We subdivide R into smaller squares with sides of length 1 and then subdivide each smaller square into two triangles. Here $m = n = 3$. We arrive at 18 triangles and four interior points (see Figure 10.7).

Let $f(x, y) = x^2 + y^4/20$ determine the boundary values. A surface S agreeing with f on the boundary will have 18 triangular patches. The area of this surface is given by Eq. (10.6), which in this case has 18 terms $\sigma_{i,j}$, $\tau_{i,j}$, indexed by $i, j = 1, 2, 3$. The sum will be a function of the heights $z_{2,2}, z_{2,3}, z_{3,2}, z_{3,3}$ of the surface

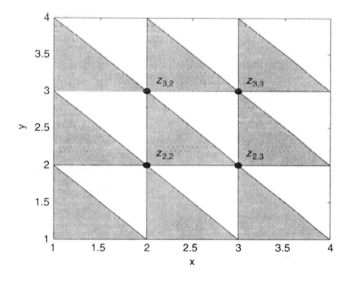

Figure 10.7 Interior points indicated with dots.

at the four interior points. Instead of the cumbersome $z(i, j)$ notation, let us set

$$a = z_{2,2}, \quad b = z_{2,3}, \quad c = z_{3,2}, \quad d = z_{3,3}.$$

Then we can think of the surface area as a function of the four variables a, b, c, d:

$$A = A(a, b, c, d).$$

The following function mfile implements this function. It uses the mfile tsurf defined earlier.

```
function out = A(a,b,c,d)
f = inline('x.^2 + (y.^4)/20', 'x', 'y')
x = [1 2 3 4]; y = [1 2 3 4];
[X,Y] = meshgrid(x,y);
Z = f(X,Y);
Z(2,2) = a; Z(2,3) = b;
Z(3,2) = c; Z(3,3) = d;
tsurf(X,Y,Z)
```

Note that the mfile replaces the values $f(2, 2)$, $f(2, 3)$, $f(3, 2)$, $f(3, 3)$ in the Z matrix with the variable values a, b, c, d, and calculates the area of the resulting triangular patch surface. In Figure 10.8 we see the triangular approximation to the original surface $z = f(x, y) = x^2 + y^4/20$ on the left. Its area is calculated to be 64.5593. The interior values are then changed to make the surface closer to linear: $a = 8.3$, $b = 13.05$, $c = 12.55$, $d = 17.05$. The resulting surface is displayed on the right. The area of the surface on the right is 60.9110.

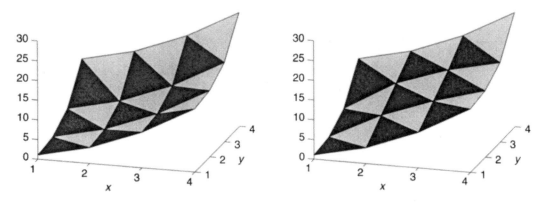

Figure 10.8 On the left, the triangular patch surface with boundary data and interior points given by $z = f(x, y) = x^2 + y^4/20$ on the square $R = [1, 4] \times [1, 4]$. On the right, the surface with modified values at the interior points.

In the preceding example we used only a small number of triangular patches to construct the surface, and there were only four interior points. The area of the surface was a function of the values at the four interior points. This discrete version of Plateau's problem becomes a problem of minimizing the area function, which depends on four variables. If we chose to make a finer mesh on R, we would increase the number of interior points. We could easily be faced with the problem of minimizing an area function that depended on 100 or more variables. Setting the derivatives of the area function equal to zero to find the minimum would yield a system of perhaps 100 nonlinear equations in 100 unknowns. Newton's method, as described in Chapter 7, can be used on such system, and there are other methods that can also be used.

A more direct, but usually slower, way to approach the problem of minimizing a function A of many variables is to use a discrete search method. Given a starting value z_0, the algorithm computes the values of $A(z)$ near z_0 and looks for a point z_1 such that $A(z_1) < A(z_0)$. Then the search is done again through points near z_1. Of course the search algorithm is made as efficient as possible.

MATLAB has a minimizing routine for functions of one variable, called `fminbnd`. It also has a routine for minimizing functions of many variables called `fminsearch`. To get information about `fminbnd` and `fminsearch`, use the `help` command. The instructor demo `minsurf` employs this procedure.

Exercises

1. Use `simp2` to estimate the area of the part of the sphere $x^2 + y^2 + z^2 = 4$ that lies over the square $-1 \leq x, y, \leq 1$. Use a sufficiently fine mesh so that the error estimate is on the order of 10^{-4} or less.

2. Let $f(x, y) = x^2 + y^3$.
    a) Graph the surface over the disk of radius 2.
    b) Set up the integral for the area of the graph of $f(x, y)$ over the set $\{|y| \leq x, \ x^2 + y^2 \leq 4\}$. Change to polar coordinates, and then estimate the integral to within 10^{-4} using `simp2`.

3. Find the area of the surface $z = \exp(-x^2 - y^2)$ over the triangle $G = \{x, y \geq 0, \ x + y \leq 1\}$.
    a) Express the surface area as a double integral over G.
    b) Make a change of variable $x = u, y = y(u, v)$ that maps the rectangle $R = \{0 \leq u \leq 1, \ 0 \leq v \leq 1\}$ onto G.
    c) Estimate the resulting double integral over R with an error on the order of 10^{-4} or less using `simp2.m`.

4. Parameterize the ellipsoid surface

$$\left(\frac{x}{a}\right)^2 + \left(\frac{y}{b}\right)^2 + \left(\frac{z}{c}\right)^2 = 1$$

using spherical coordinates,

$$x = a\cos\theta\sin\phi, \qquad y = b\sin\theta\sin\phi, \qquad z = c\cos\phi.$$

a) Express the surface area of the ellipsoid as a double integral over the rectangle in θ, ϕ space, $R = \{0 \le \theta \le 2\pi, \ 0 \le \phi \le \pi\}$.

b) Use `simp2` to estimate the integral for $a = 1, b = 2, c = 3$ with an error on the order of 10^{-4} or less.

★   **5.** Recall the boat hull described in Exercise 9 of Chapter 4. The curved piece is the graph of $y = (2x/a(z))^2$ on the set $G = \{-a(z) \le x \le a(z), \ 0 \le z \le 20\}$. The function $a(z) = -.0166z^2 + .2245z + 2.25$.

a) Write the area of the curved piece of the hull as a double integral with respect to x and z over the set G.

b) Make the change of variable $x = a(v)u, \ z = v$ to yield a double integral over the rectangle $R = \{-1 \le u \le 1, \ 0 \le v \le 20\}$.

c) Use `simp2` to estimate the integral with an error on the order of 10^{-4} or less.

d) Find the total area of the boat hull by adding on the area of the flat piece that is the stern.

★   **6.** Let the surface S be given by $z = x^2 + y^3$ over the rectangle $0 \le x \le 3, \ 0 \le y \le 2$.

a) Graph the surface using `surf`.

b) Assuming a constant density $\delta = .55$ gms/cm^2, make a numerical estimate of the mass of the sheet of metal forming the surface S. Use `simp2.m`.

c) Make numerical estimates of the coordinates of the center of mass of the metal sheet, given by

$$\bar{x} = \frac{\delta}{m}\int\int_S x\,dS, \qquad \bar{y} = \frac{\delta}{m}\int\int_S y\,dS, \qquad \bar{z} = \frac{\delta}{m}\int\int_S z\,dS.$$

7. A pagoda-type roof over a circular room is a surface of revolution that is obtained by revolving the curve $x = -.12z^3 + 5.4z^2 - 82z + 420, \ 10 \le z \le 15$ about the z axis.

a) Graph the surface.

b) Use Eq. (10.5) to express the surface area of this roof as a one-dimensional integral over the interval $10 \le v \le 15$.

c) Estimate the integral using `quadl`.

8. Let a body of water be h feet deep. The hydrostatic pressure, in pounds per square foot, increases as the depth below the surface increases. In fact, the pressure p is given by the formula

$$p(z) = 62.5(h - z),$$

where z is the height above the bottom, $0 \leq z \leq h$.

An empty vessel sits on the bottom. It is bounded by a closed surface that consists of a paraboloid, $z = h/4 - x^2 - y^2$, and the flat bottom $z = 0$.

a) Write an integral expression for the total hydrostatic force on the curved surface of the vessel.

b) Evaluate the integral for $h = 100$ feet.

★ **9.** Let S be the torus, given parametrically by

$$
\begin{aligned}
x &= (100 + 2\cos v)\cos u \\
y &= (100 + 2\cos v)\sin u \\
z &= 100 + 2\sin v, \qquad 0 \leq u, v \leq 2\pi
\end{aligned}
$$

Let $\rho(x, y, z) = \sqrt{x^2 + y^2}$ be the charge density on S in coulombs per unit area. Find the total charge on S.

★ **10.** Let $f(x, y) = x \exp(-x^2 - 2y^2)$, and let R be the rectangle $\{0 \leq x, y \leq 1\}$.

a) Express the surface area of the graph of f over R as a double integral over R. Use simp2 to estimate the integral with an error on the order of 10^{-4}. You should use an mfile for the integrand, because it is too complicated to do as an inline function.

b) Now use the mfile tsurf.m to estimate the surface area using triangles. Take $m = n$. How large must n be so that the area calculated by tsurf is within .05 of the value of the surface area as estimated by simp2? To use tsurf you can write f as an inline function.

11. Suppose the coordinates of points on a surface S are determined by some numerical procedure. The surface is over a rectangle $R = [0, 1] \times [0, 1]$ with meshpoints $x_j = (j - 1)/10$, $j = 1, \ldots 11$, and $y_i = (i - 1)/10$, $i = 1, \ldots, 11$. The heights of the surface at each meshpoint are tabulated in an 11×11 matrix Z, with $Z(i, j)$ being the height at (x_j, y_i).

a) If you are given the matrix Z, how would you approximate the area of the surface S?

b) Suppose the elements of Z are random numbers lying in the interval $0 \leq z \leq .1$. The MATLAB command P = .1* rand(11,11) generates such a matrix.

Let the surface S be at a constant height $z = 1$ with a random perturbation at the meshpoints. Use $z = 1+P$. Find the area of the surface using `tsurf`. Do it several times. The result should be different each time, because the random numbers are different.

★   **12.** Let R be the rectangle $[0, 2] \times [0, 4]$. Let S be the surface over R consisting of four triangles with common vertex at $(1, 2)$. See Figure 10.9. The heights of the surface at the corners are

$$
\begin{array}{llll}
(0,0) & \text{height} = 1 & \quad (2,0) & \text{height} = 3 \\
(0,4) & \text{height} = 2 & \quad (2,4) & \text{height} = 1.
\end{array}
$$

The height z of the surface at $(1, 2)$ is to be determined.

    a) Write a sum of four terms that expresses the area A of the surface S as a function of z, $A(z)$.

    b) Write an mfile for $A(z)$.

    c) Plot $A(z)$ on the interval $0 \le z \le 3$. Estimate by eye the value of z that minimizes the surface area A.

    d) Use the MATLAB routine `fminbnd` to find the value of z that minimizes $A(z)$, and then calculate the corresponding value of A.

★   **13.** Let values h_1, h_2, h_3, h_3 be prescribed at the vertices $\mathbf{p}_1 = (2, 0)$, $\mathbf{p}_2 := (0, 1)$, $\mathbf{p}_3 = (-2, 0)$, and $\mathbf{p}_4 = (0, -1)$. The values z at $\mathbf{p}_5 = (-1, 0)$ and w at $\mathbf{p}_6 := (1, 0)$ are to be chosen to make the surface as small as possible. See Figure 10.10.

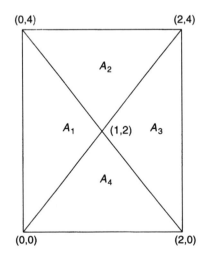

Figure 10.9 Triangles for Exercise 12.

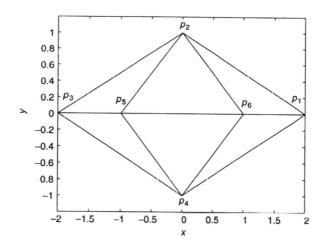

Figure 10.10 Location of vertices for Exercise 13.

a) Write a sum of six terms that expresses the surface area A as a function $A(z, w)$. Let $h_1 = 1$, $h_2 = 2$, $h_3 = 4$, and $h_4 = 3$.

b) Write an mfile A.m for $A(z, w)$ and graph A on the square $1 \le z, w \le 4$. Plot the contours and use the mfile findcrit to estimate the values of z, w that minimize A.

c) We cannot use tsurf to graph the surface found in part b) because the points $\mathbf{p}_j$ do not lie on rectangular grid. To display this surface, you must write a script file that uses the command fill3 to fill in each of the six triangular elements of the surface.

14. Suppose a surface S is given in the form of a matrix Z of heights over a rectangle $R = [a, b] \times [c, d]$. We wish to estimate $\int \int_S \delta \, dS$.

Let meshpoints (x_j, y_i) be given in R with $a = x_1 < x_2 < \cdots < x_n = b$ and $c = y_1 < y_2 < \cdots < y_m = d$. Let $z_{i,j}$ be the height of the surface over the point (x_j, y_i). Using Eq. (10.6) as a guide, write a formula to approximate $\int \int_S \delta \, dS$. Evaluate the function δ at the centroid of each triangular surface element. If $\mathbf{p}_1$, $\mathbf{p}_2$, $\mathbf{p}_3$ are points in R^3, the centroid of the triangle with these vertices is given by $\mathbf{c} = (\mathbf{p}_1 + \mathbf{p}_2 + \mathbf{p}_3)/3$.

Write a MATLAB script that implements this procedure over the square $R = 0 \le x, y \le 1$ for various values of m and n. Let the surface S be given by $z = f(x, y) = x^2 + y^3$. Let $\delta(x, y, z) = \sqrt{x^2 + y^2 + z^2}$.

11

Integrals of Vector Fields Over Curves and Surfaces

Prepared mfiles used in this chapter

```
lint    flux2    curl
```

11.1 Vector fields

A two-dimensional vector field is written in Cartesian coordinates with two functions, $u(x, y)$ and $v(x, y)$. At each point (x, y) in a set G, we attach the vector $\mathbf{F} = [u(x, y), v(x, y)]$. Vector fields in two-dimensional space are easily displayed with the command `quiver`. Recall from Chapter 5 that first we need to write the functions u and v either as mfiles or as inline functions. Then we need to construct a meshgrid [X,Y] over some rectangle R. Finally, the call is `quiver(X,Y,u(X,Y),v(X,Y))`.

Example 11.1

We display the vector field $\mathbf{F} = [1, x + y^2]$ over the rectangle $R = \{-2 \leq x \leq 3, -1 \leq y \leq 2\}$. You should be careful not to put too many arrows in the figure, making it difficult to interpret. A mesh with between 10 and 15 subdivisions in each direction is usually fine enough. See Figure 11.1.

```
>> u = inline('0*x +1', 'x', 'y')
>> v = inline('x + y.^2','x', 'y')
>> x = linspace(-2,3,11);
>> y = linspace(-1,2,11);
```

219

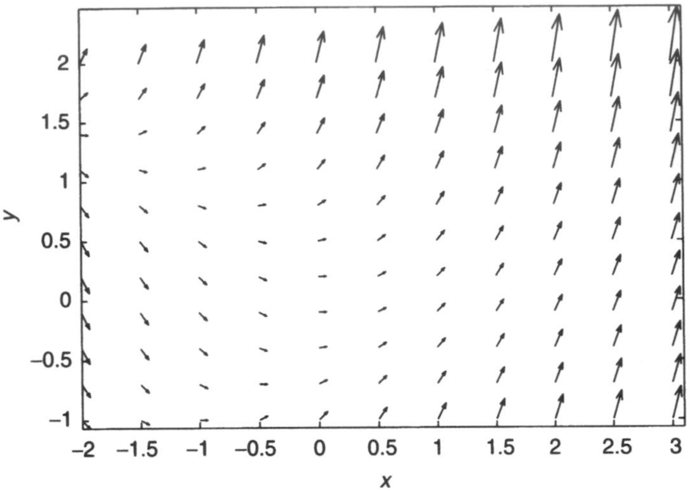

Figure 11.1 Vector field $\mathbf{F} = [1, x + y^2]$.

```
>> [X,Y] = meshgrid(x,y);
>> U = u(X,Y);   V = v(X,Y);
>> quiver(X,Y,U,V)
>> axis image
```

Notice that the definition of u as an inline function lists both x and y as variables, even though u is actually a constant function.

The last command, axis image, sets the tick marks the same on both axes (as does the command axis equal), but in addition does not leave extra space at the edges of the plot, where the arrows may hang over. Notice also that the length of the arrows is scaled to represent the relative magnitude of the vectors. If there is some place in the plotting rectangle where the vector field is quite large, it may appear to be zero at points where the magnitude is more moderate. To see a unit vector field displayed with the same directions as $\mathbf{F}$, set

```
>> U1 = U./sqrt(U.^2 + V.^2)
>> V1 = V./sqrt(U.^2 + V.^2).
>> quiver(X,Y,U1,V1)
>> axis image
```

The result is shown in Figure 11.2. Compare with Figure 11.1.

Vector fields in three-dimensional space have three component functions, $\mathbf{F} = [u(x, y, z),\ v(x, y, z),\ w(x, y, z)]$, and attach this vector to each point (x, y, z) in some set G in xyz space. MATLAB also has a way of displaying three-dimensional vector fields. First we make mfiles or inline functions for u, v, and

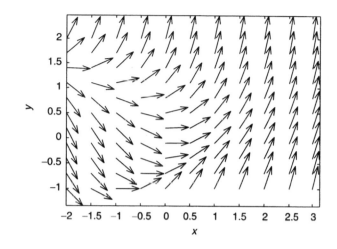

Figure 11.2 Normalized vector field F/‖F‖.

w. The MATLAB graphing command `quiver3` displays the arrows of **F** attached to any two-dimensional surface S that lies inside a three-dimensional rectangular space R. Recall that we used `quiver3` in Example 6.4 to display the gradient vectors ∇f attached to a level surface of f. To fill a three-dimensional rectangular space R with arrows, we attach the arrows to a family of surfaces, for example, planes $z = $ constant. This can be done with a short loop in a script mfile.

Example 11.2

Consider the vector field

$$\mathbf{F}(x, y, z) = [1, \ x + y^2, \ z].$$

We shall display this vector field in the rectangle

$$R = \{-2 \le x \le 3, \ -1 \le y \le 2, \ -1 \le z \le 1\}.$$

Here is a script mfile to do this. See Figure 11.3.

```
u = inline('1 + 0*x', 'x', 'y', 'z')
v = inline('x + y.^2', 'x', 'y', 'z')
w = inline('z', 'x', 'y', 'z')
x = linspace(-2,3,6);
y = linspace(-1,2,6);
[X,Y] = meshgrid(x,y);
for z = -1:.4:1
```

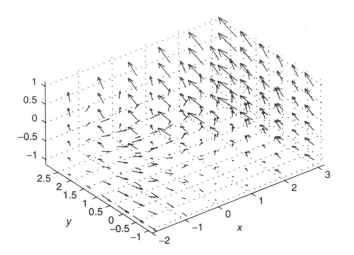

Figure 11.3 Three-dimensional plot of the vector field $\mathbf{F} = [1, \ x + y^2, \ z]$.

```
Z  =  z+0*X;
U  =  u(X,Y,Z);
V  =  v(X,Y,Z);
W  =  w(X,Y,Z);
quiver3(X,Y,Z,U,V,W)
hold on
      end
```

Again notice that the inline definitions of u, v, and w list all three variables, x, y, z. We have taken the xy mesh even coarser than in Example 11.1, and we are using only six levels in the z direction. To see the vector field properly, it is usually necessary to view it from different angles using the `rotate3d` command. The call for `quiver3` is the exact analog of that for `quiver`, but with three arguments for the point of attachment and three arguments for the components of the vector field. In versions 5.3 and higher of MATLAB, `quiver3` can take three-dimensional arrays as arguments. In this case the loop can be omitted and the execution is a bit faster. After `u,v,w,x,y` are defined, use the commands

```
z = linspace(-1,1,6);
[X,Y,Z] = meshgrid(x,y,z);
U = u(X,Y,Z);
V = v(X,Y,Z);
W = w(X,Y,Z);
quiver3(X,Y,Z,U,V,W)
```

11.2   Line integrals

A line integral of a vector field $\mathbf{F} = [u, v, w]$ along an oriented curve C from P to Q is an integral written

$$\int_C \mathbf{F} \cdot d\mathbf{r} = \int_C u\,dx + v\,dy + w\,dz. \tag{11.1}$$

It is defined as a limit of sums

$$\int_C \mathbf{F} \cdot d\mathbf{r} = \lim_{n \to \infty} \sum_{i=1}^{n} \mathbf{F}(P_i) \cdot (P_{i+1} - P_i),$$

where P_i, $i = 1, \ldots, n + 1$, is a sequence of points on C, with $P_1 = P$ the initial point and $P_{n+1} = Q$ the terminal point. See Figure 11.4. From the definition, it follows that the line integral

$$\int_C \mathbf{F} \cdot d\mathbf{r} = \int_C \mathbf{F} \cdot \mathbf{T}\, ds,$$

where $\mathbf{T}$ is the unit tangent vector to the curve. Hence line integral (11.1) is the integral with respect to arc length of the tangential component of $\mathbf{F}$ and is interpreted

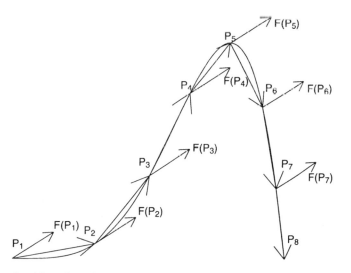

Figure 11.4 Curve C with points P_i, vectors $P_{i+1} - P_i$, and vector field $\mathbf{F} = [u, v]$, with $u(x, y) \equiv$ .15 and $v(x, y) \equiv .1$.

as the work done by the force field $\mathbf{F}$ on a body following the curve C in the given orientation.

When the curve C is parameterized by a continuously differentiable function $\mathbf{r}(t)$ on the interval $a \le t \le b$, with $\mathbf{r}(a) = P$ and $\mathbf{r}(b) = Q$, the line integral can be written

$$\int_C \mathbf{F} \cdot d\mathbf{r} = \int_a^b \mathbf{F}(\mathbf{r}(t)) \cdot \mathbf{r}'(t) \, dt. \tag{11.2}$$

In this form the line integral may be evaluated either numerically or symbolically.

Example 11.3

We use Simpson's rule to estimate the value of the line integral $\int_C \mathbf{F} \cdot d\mathbf{r}$, where

$$\mathbf{F}(x, y, z) = [x, z, \exp(x + y)]$$

and C is parameterized by $\mathbf{r}(t) = [t \cos t, \ t \sin t, \ t^2], \ 0 \le t \le 2\pi$. A call is made to the mfile `simpvec`, written in Section 9.2.

```
% Define the components of F.
u = inline('x', 'x', 'y', 'z')
v = inline('z', 'x', 'y', 'z')
w = inline('exp(x+y)', 'x', 'y', 'z')

% Chose the t values for Simpson's rule.
n = 200;
t = linspace(0, 2*pi, n+1); dt = 2*pi/n;
s = simpvec(n);

% Calculate the x,y,z values along the curve.
x = t.*cos(t); y = t.*sin(t); z = t.^2;

% Calculate the components of rdot along the curve.
xdot = cos(t) -t.*sin(t);
ydot = sin(t) +t.*cos(t);
zdot = 2*t;

% Calculate the terms of the integrand
I1 = u(x,y,z).*xdot;
I2 = v(x,y,z).*ydot;
I3 = w(x,y,z).*zdot;

% Compute the integral using Simpson.
integral = dot(s,(I1 +I2 +I3))*dt/3;
```

However, if a parameterization of C is not available, we can make an estimate of line integral (11.1) as follows. Approximate the curve C by a sequence of line segments between the points P_j on the curve, $j = 1, \ldots, n+1$, as in Figure 11.4. The line segment L_j from P_j to P_{j+1} can be parameterized by

$$\mathbf{r}_j(t) = (1 - t)P_j + t P_{j+1}, \qquad 0 \le t \le 1,$$

with $\mathbf{r}'_j = P_{j+1} - P_j$. Then

$$\int_C \mathbf{F} \cdot d\mathbf{r} \approx \sum_{j=1}^{n} \int_{L_j} \mathbf{F} \cdot d\mathbf{r}$$

$$= \sum_{j=1}^{n} \int_0^1 \mathbf{F}(\mathbf{r}_j(t)) \cdot (P_{j+1} - P_j)\, dt$$

$$= \sum_j (x_{j+1} - x_j) \int_0^1 u(\mathbf{r}_j(t))\, dt$$

$$+ \sum_j (y_{j+1} - y_j) \int_0^1 v(\mathbf{r}_j(t))\, dt$$

$$+ \sum_j (z_{j+1} - z_j) \int_0^1 w(\mathbf{r}_j(t))\, dt.$$

Each of the integrals can be done either symbolically or numerically using Simpson's rule or `quadl`.

Example 11.4

Suppose that a curve C in the xy plane has been found numerically as the trajectory of a particle. The coordinates of the points on C are as given in the following table:

x	0	.1	.25	.4	.54	.76	.82	.93	1
y	0	.005	.0312	.0800	.1458	.2888	.3362	.4352	.5

Let a force field be given by $\mathbf{F}(x, y) = [x \cos y, x+y]$. We want to estimate the work done on the particle by $\mathbf{F}$ as it follows curve C from $P_1 = (0, 0)$ to $P_9 = (1, .5)$. This is done in the following script. We use Simpson's rule, with $n = 50$ on each segment.

```
x = [0 .1 .25 .4 .54 .76 .82 .93 1]
y = [0 .005 .0312 .0800 .1458 .2888 .3362 .4352 .5]

% Define the components of F.
u = inline('x.*cos(y)', 'x', 'y')
v = inline('x+y', 'x', 'y')

% Prepare the Simpson vector.
n = 50;
s = simpvec(n);
dt = 1/n;

% Initialize Work.
Work = 0;
for j = 1:8
    % Calculate the values of x and y along the
    % jth segment.
    xx = linspace(x(j), x(j+1), n+1);
    yy = linspace(y(j), y(j+1), n+1);

    % Calculate the integrand.
    I = (x(j+1) - x(j))*u(xx,yy)
            + (y(j+1) - y(j))*v(xx,yy)

    % Calculate the jth integral.
    integral = dot(s,I)*dt/3;
    Work = Work + integral
end
Work = .9353
```

mfile lint

The mfile lint.m uses the procedure of Example 11.4 to estimate the line integral of a two-dimensional vector field $\mathbf{F} = [u(x, y), v(x, u)]$ along a polygonal path determined by the user by clicking on the figure. The call is lint(u,v,corners), where, as usual, corners is a vector $[a, b, c, d]$ that defines a rectangle R. Remember to use single quotes in the call when u and v are given in mfiles. You are then asked to enter the number N of line segments in the path. The file uses quiver to plot the vector field over R. Then click on the figure with the left mouse button to start the path of integration. A second click produces a line from the first point to the second point and computes the work done by the vector field along this line. This procedure can be repeated a total of N times. The cumulative value of the line integral is shown on the screen. The program also calculates the line integral

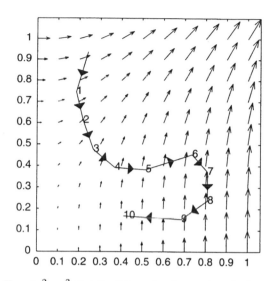

Figure 11.5 Vector field $\mathbf{F} = [y^2, \ x^3/6 + 3x/2]$, and polygonal path determined by mfile `lint.m`.

$\int_C x\,dy$. When the curve C is closed and forms the positively oriented boundary of a polygon, this line integral is the area of the polygon. A sample figure window is shown in Figure 11.5 with $N = 10$.

A vector field is *conservative* if it is the gradient of a scalar potential, $\mathbf{F}(x, y) = \nabla f$. In component terms, $u(x, y) = f_x(x, y)$ and $v(x, y) = f_y(x, y)$. If the vector field $\mathbf{F}$ is conservative, the line integrals $\int_C \mathbf{F} \cdot ds$ are independent of path. Equivalently, $\int_C \mathbf{F} \cdot ds = 0$ for every closed curve such that the components of $\mathbf{F}$ are continuously differentiable in the interior of C. In the exercises we shall use `lint.m` to explore examples of conservative and nonconservative vector fields.

11.3   Curl and Green's theorem

The curl of a three-dimensional vector field

$$\mathbf{F}(x, y, z) = [u(x, y, z), \ v(x, y, z), \ w(x, y, z)],$$

where u, v, w are continuously differentiable functions, is defined to be the vector field

$$\mathrm{curl}(\mathbf{F}) = [w_y - v_z, \ u_z - w_x, \ v_x - u_y].$$

Green's theorem helps to give a physical and geometrical meaning to this quantity. First we restrict our discussion to a vector field that has only two nonzero

Integrals of Vector Fields Over Curves and Surfaces

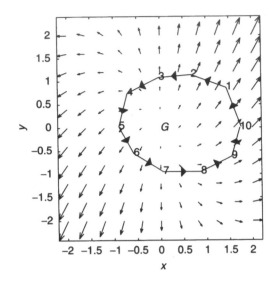

Figure 11.6 Line integral around the boundary of set G with positive orientation.

components that depend only on x, y,

$$\mathbf{F} = [u(x, y), \ v(x, y), \ 0].$$

In this case $\mathrm{curl}(\mathbf{F}) = [0, 0, v_x - u_y]$. Now we assume this vector field is defined in a set G in xy space with piecewise smooth boundary C. We also assume that C has the positive orientation with respect to G. In particular, the outer boundary of G is oriented in the counterclockwise direction. See Figure 11.6. Then we have

Green's theorem

$$\int_C u\,dx + v\,dy = \int\int_G (v_x - u_y)\, dA(x, y), \tag{11.3}$$

which may also be stated,

$$\int_C \mathbf{F} \cdot \mathbf{T}\, ds = \int_C \mathbf{F} \cdot d\mathbf{r} = \int\int_G \mathrm{curl}(\mathbf{F}) \cdot \mathbf{k}\, dA(x, y). \tag{11.4}$$

The line integral on the left of Eq. (11.3) represents the work done by a force field $\mathbf{F}$ in circling around the closed path C. In fluid flow, $\mathbf{F} = [u, v]$ is the velocity field of the fluid, with u being the x component of velocity and v being the y component. In this context, the line integral is called the *circulation* of the fluid around the closed path C. On the right-hand side, $\mathbf{k}$ is the vector $[0, 0, 1]$.

Green's theorem gives meaning to the curl in the following way. Consider a small square of side h centered at (x_0, y_0). Applying Green's theorem to this small square, which we call R_h, and dividing both sides of Eq. (11.3) by h^2, we have

$$\frac{1}{h^2} \int_C u\, dx + v\, dy \;=\; \frac{1}{h^2} \int\int_{R_h} (v_x - u_y)\, dA(x, y)$$

$$= \; \text{average over } R_h \text{ of } (v_x - u_y).$$

If we assume curl(**F**) is continous, then the right-hand side of this equation converges to $v_x(x_0, y_0) - u_y(x_0, y_0)$ as $h \to 0$. Thus curl(**F**)(x_0, y_0) is the limit of the circulation per unit area around smaller and smaller squares. Hence we can interpret curl(**F**)(x_0, y_0) as the circulation per unit area at the point (x_0, y_0).

Example 11.5

We can use the mfile `lint.m` to illustrate this last idea. Let the vector field **F** = $[1 - y^2, 0]$ in the region $-1 \le y \le 1$. This vector field represents the two-dimensional flow of a viscous fluid from left to right in a channel whose walls are the lines $y = \pm 1$. The fluid sticks to the walls (the no-slip boundary condition). We see immediately that curl(**F**) $= 2y\mathbf{k}$. Now we use the mfile `lint.m` to calculate the circulation around smaller and smaller paths centered at the point $(0, 1/2)$, where curl(**F**) $\cdot \mathbf{k} = 1$. The program also computes the areas of the enclosed polygon. We first take corners = [-1 1 -1 1] and calculate the circulation around a triangle with vertices at approximately $P_1 = (-.5, 0)$, $P_2 = (.5, 0)$, $P_3 = (0, 1)$. Dividing the circulation by the area, we obtain .6747 (the exact answer is 2/3). Repeating the calculation with a triangle with vertices at approximately $P_1 = (-.25, .3)$, $P_2 = (.25, .3)$, $P_3 = (0, .7)$, and again dividing the circulation by the area, we obtain .8655 (the exact answer is .86666...). The smaller of these two triangles is shown in the upper part of Figure 11.7. Since it is hard to close the path by eye with the cursor, to obtain better accuracy we should graph the vector field over a smaller rectangle enclosing the point $(0, 1/2)$ and repeat the calculations over triangles centered at $(0, 1/2)$. As expected, we see that the value of the circulation, divided by the area enclosed by the path, approaches 1. We have used triangles here, but any closed polygons could be used.

In the lower half of Figure 11.7, we see another triangle. Because of the way **F** is changing, i.e., u is increasing for $y < 0$, the circulation around this triangle is negative, and curl(**F**) points in the $-\mathbf{k}$ direction.

In general, it is not easy to see if a vector field has a nonzero curl by looking at a graph of the vector field. Nevertheless, we can present some examples of typical two-dimensional vector fields whose curl can be determined visually.

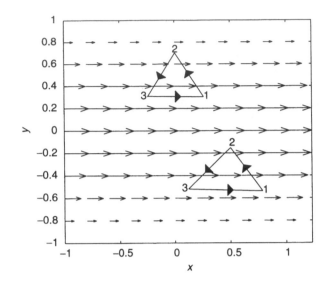

Figure 11.7 Circulation around triangles of the vector field $\mathbf{F}(x, y) = [1 - y^2, 0]$.

Example 11.6

A *shear flow* is defined as follows. Let $\mathbf{p} = [p_1, p_2]$ be a constant vector, and let $a(s)$ be a continuously differentiable function of one variable. Then the function $(x, y) \to a(p_2 x - p_1 y)$ is constant on lines parallel to $\mathbf{p}$. The gradient

$$[\partial_x, \partial_y] a(p_2 x - p_1 y) = a'(p_2 x - p_1 y)\mathbf{q}$$

is always parallel to the vector $\mathbf{q} = [p_2, -p_1]$. Let the vector field $\mathbf{F} = a(p_2 x - p_1 y)\mathbf{p}$. Then the vector field $\mathbf{F}$ is always parallel to $\mathbf{p}$, and it changes only in the direction $\mathbf{q}$, which is orthogonal to $\mathbf{p}$. We have

$$\text{curl}(\mathbf{F}) = (v_x(x, y) - u_y(x, y))\mathbf{k} = a'(p_2 x - p_1 y)(p_1^2 + p_2^2)\mathbf{k}.$$

Hence curl($\mathbf{F}$) is nonzero whenever $a' \neq 0$. The vector field of Example 11.5 is a shear flow with $\mathbf{p} = \mathbf{i}$.curl($\mathbf{F}$) $= 2y\mathbf{k}$ points in the positive direction for $y > 0$ and in the negative $\mathbf{k}$ direction for $y < 0$.

Example 11.7

Vortex flow: Let

$$\mathbf{F}(x, y) = \left[\frac{-y}{r}, \frac{x}{r}\right],$$

where $r = \sqrt{x^2 + y^2}$. **F** is a unit vector that describes a fluid rotating in the counterclockwise direction about the origin. We have

$$\text{curl}(\mathbf{F})(x, y) = (v_x(x, y) - u_y(x, y))\mathbf{k} = \left(\frac{1}{r}\right)\mathbf{k}.$$

This vector field is singular at the origin because the curl becomes infinite there.

mfile curl

This mfile can help you determine where a two-dimensional vector field has a nonzero curl and indicates the sign and magnitude by color. The call is `curl(u,v,corners)`, where, as usual, $u(x, y)$ and $v(x, y)$ are given in mfiles or as inline functions. Remember to use single quotes on u and v in the call when they are given in mfiles. `corners` is the vector of corner coordinates of the graphing rectangle. When the call is made, the vector field is graphed, and curl(**F**) is computed using difference quotient approximations to the derivatives. The vector field is superimposed on a pcolor plot of curl(**F**). A color bar is displayed on the right of the figure to indicate how the color scale relates to the values of curl(**F**). See Figure 11.8.

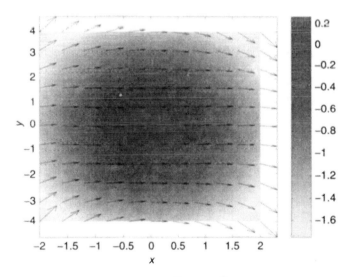

Figure 11.8 Vector field $\mathbf{F}(x, y) = (1/4)[1, x - x^3/3 - xy^2/4]$ displayed together with a pcolor plot of curl(**F**).

11.4   Flux integrals

Let $\mathbf{F}(x, y, z) = [u(x, y, z), v(x, y, z), w(x, y, z)]$ be a vector field on some region of xyz space, and let S be a piece of two-dimensional surface in that region. We assume that S is an orientable surface. This means that at each point $(x, y, z) \in S$, there is defined a unit normal vector $\mathbf{n}$ that varies continuously as (x, y, z) moves over S. Then the *flux integral* of $\mathbf{F}$ over S is

$$\int\int_S \mathbf{F} \cdot \mathbf{n} \, dS, \tag{11.5}$$

where dS is the element of surface area on S. When $\mathbf{F}$ is the velocity of a fluid, expression (11.5) is the volume of fluid crossing the surface S per unit time. When the flux is positive, the net flow across S is in the direction specified by $\mathbf{n}$.

For a vector field $\mathbf{F}(x, y) = [u(x, y), v(x, y)]$ and a curve C with normal vector $\mathbf{n}$, the flux integral is

$$\int_C \mathbf{F} \cdot \mathbf{n} \, ds, \tag{11.6}$$

where ds is the element of arc length on C. In particular, if C is the level curve of a function $g(x, y)$, then we may take

$$\mathbf{n} = \pm \frac{\nabla g}{\|\nabla g\|} = \pm \frac{[g_x, \ g_y]}{\sqrt{g_x^2 + g_y^2}}.$$

If, instead, C is parameterized by a piecewise smooth function $\mathbf{r}(t) = (x(t), y(t))$, we may take

$$\mathbf{n} = \pm \frac{[y'(t), -x'(t)]}{[x'(t)^2 + y'(t)^2]^{1/2}}.$$

In this case, the flux integral (11.6) reduces to

$$\int_C \mathbf{F} \cdot \mathbf{n} \, ds = \pm \int_C u \, dy - v \, dx.$$

mfile flux2

The mfile `flux2.m` computes the flux of a two-dimensional vector field through a boundary composed of line segments. It is used in the same manner as the mfile `lint.m`. The call is `flux2(u,v,corners)` where the components u, v of $\mathbf{F}$ are given in mfiles or as inline functions. Remember to use single quotes in the call when u and v are given in mfiles. corners is the vector of corner coordinates $[a, b, c, d]$. After the call, enter the number N of segments in the path. You begin the path of integration with a first click on the figure. This time, after the second click, the program computes the flux $\int \mathbf{F} \cdot \mathbf{n} \, ds$ along the segment. The direction of the normal is shown by the red arrow. This can be done for N segments.

Example 11.8

Let the vector field be $\mathbf{F}(x, y) = [u, v] = [x^2/2, y^2/2]$. We are going to find the flux through the boundaries of two triangles contained in the square $-2 \leq x, y \leq 2$. The first triangle is in the upper right of Figure 11.9, and the second triangle is in the lower left of the same figure.

```
>> u = inline('.5*x.^2', 'x', 'y')
>> v = inline('.5*y.^2', 'x', 'y')
>> corners = [-2 2 -2 2]
>> flux2(u,v,corners)
   Enter the number of segments 3
flux =
      1.2852

>> hold on
>> flux2(u,v,corners)
   Enter the number of segments 3
flux = -9.885
```

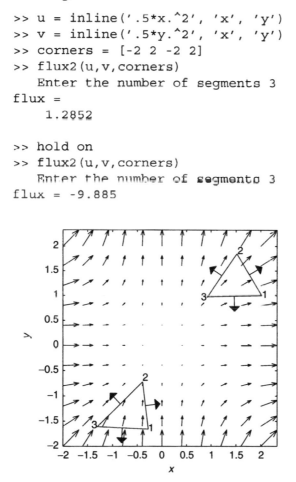

Figure 11.9 Two flux calculations for the vector field $\mathbf{F} = [x^2/2, y^2/2]$.

11.5 The divergence theorem

Let $\mathbf{F}(x, y, z) = [u(x, y, z), v(x, y, z), w(x, y, z)]$ be a vector field. The divergence of $\mathbf{F}$ is the scalar function

$$\mathrm{div}(\mathbf{F}) = u_x(x, y, z) + v_y(x, y, z) + w_z(x, y, z).$$

The Gauss divergence theorem connects the integral of $\mathrm{div}(\mathbf{F})$ with a flux integral. Let G be a bounded set in xyz space, which is bounded by an orientable surface S. Let $\mathbf{n}$ denote the unit exterior normal to S.

Divergence theorem

$$\int\int\int_G \mathrm{div}(\mathbf{F}) \, dV(x, y, z) = \int\int_S \mathbf{F} \cdot \mathbf{n} \, dS.$$

The divergence theorem gives a meaning to $\mathrm{div}(\mathbf{F})$ in the following way. Let R_h be a small cube with side h centered at the point (x_0, y_0, z_0). Then applying the divergence theorem over R_h and dividing by the volume h^3, we have

$$\frac{1}{h^3} \int\int_S \mathbf{F} \cdot \mathbf{n} \, dS = \frac{1}{h^3} \int\int\int_{R_h} \mathrm{div}(\mathbf{F}) \, dV. \tag{11.7}$$

The right-hand side of Eq. (11.7) is the average of $\mathrm{div}(\mathbf{F})$ over the cube R_h. If the components of $\mathbf{F}$ are continuously differentiable, this average converges to $\mathrm{div}(\mathbf{F})(x_0, y_0, z_0)$ as $h \to 0$. Hence for small h we can say

$$\mathrm{div}(\mathbf{F})(x_0, y_0, z_0) \approx \frac{1}{h^3} \int\int_S \mathbf{F} \cdot \mathbf{n} \, dS.$$

Thus $\mathrm{div}(\mathbf{F})$ has the units of flux/unit volume. In the context of fluid flow, $\mathbf{F}$ is the velocity vector of the fluid, with u being the x component, v being the y component, and w being the z component. Since velocity has the units of length/time, the flux has the units of volume/time. It is the volume of fluid that flows across the boundary surface S per unit time. Hence in this context, $\mathrm{div}(\mathbf{F})$ is the volume/time of fluid that is being created or absorbed per unit volume.

Of course, there is a two-dimensional version of the divergence theorem. The mfile `flux2.m` can help us understand the divergence in the two-dimensional case.

Example 11.9

Let $\mathbf{F}(x, y) = [x^2/2, \sin(\pi y/2)]$. We shall compute the flux and divide by the area of the enclosed polygon to get an approximation to $\operatorname{div}(\mathbf{F})(x, y) = u_x(x, y) + v_y(x, y)$. See Figure 11.10. The square in the lower left of the figure has vertices at approximately $(-1.5, -1.5)$, $(-.5, -1.5)$, $(-.5, -.5)$, $(-1.5, -.5)$. The calculated flux is -1.0212, and the calculated area (which should be exactly 1) is 1.0096. Hence flux/area $= -1.0115$. The exact value of the divergence at the center of this square is -1.

The larger square has vertices at $(.5, -.5)$, $(2, -.5)$, $(2, 1)$, $(.5, 2)$. The flux is calculated to be 5.2746 and the area is calculated to be 2.2334. Here flux/area $=$ 2.3617. The smaller square inside the larger square has vertices at $(1, 0)$, $(1.5, 0)$, $(1.5, .5)$, $(1, .5)$. The calculated flux is .6666 and the calculated area is .2377. In the smaller square, flux/area $=$ 2.7766. The exact value at the center of square, $(1.25, .25)$, is $\operatorname{div}(\mathbf{F})(1.25, .25) = 1.25 + (\pi/2)\cos(\pi/8) = 2.7012$.

Generally speaking, it is difficult to determine visually the sign of the divergence of a vector field. Our intuitive feeling is that if the arrows are spreading apart, the divergence should be positive, and if they are converging, the divergence should be negative. This appears to be the case in Figure 11.10. However, the matter is more subtle, as shown in the following examples.

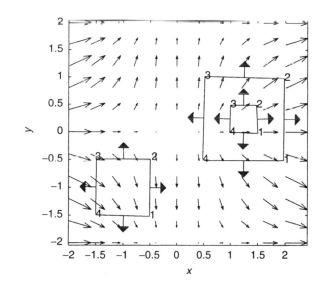

Figure 11.10 Flux calculations around squares for the vector field $\mathbf{F}(x, y) = [x^2/2, \sin(\pi y/2)]$.

Example 11.10

We define *accelerated flow* as follows: Let $\mathbf{p} = [p_1, p_2]$ be a constant vector, which will be the direction of the flow, and let $a(s)$ be a continuously differentiable function of one variable. The function $(x, y) \rightarrow a(p_1 x + p_2 y)$ is constant on lines perpendicular to $\mathbf{p}$, and the gradient of $a(p_1 x + p_2 y)$ is parallel to $\mathbf{p}$. Now we set $\mathbf{F}(x, y) = a(p_1 x + p_2 y)\mathbf{p}$. We have

$$\mathrm{div}(\mathbf{F})(x, y) = a'(p_1 x + p_2 y)(p_1^2 + p_2^2).$$

Hence $\mathrm{div}(\mathbf{F})$ is positive or negative depending on the sign of a'. Here all the vectors are parallel; they do not spread apart or converge. However, the flow speeds up or slows down depending on a, which is the reason for the term accelerated flow. A simple example with $\mathbf{p} = \mathbf{i}$ is

$$\mathbf{F}(x, y) = [x^2, \ 0].$$

We have $\mathrm{div}(\mathbf{F}) = 2x$, which is positive for $x > 0$ and negative for $x < 0$. Compare with shear flow, discussed in Example 11.6.

Example 11.11

Radial flow: We consider the vector field

$$\mathbf{F}(x, y) = \left[\frac{x}{r}, \frac{y}{r}\right],$$

where $r = \sqrt{x^2 + y^2}$. $\mathbf{F}$ is a unit vector field. We have

$$\mathrm{div}(\mathbf{F})(x, y) = \frac{1}{r}.$$

The vector field is singular at the origin because $\mathrm{div}(\mathbf{F})$ becomes infinite there. The positive divergence of this vector field is due to the fact that the arrows are spreading apart. However, as we shall see in the exercises, if the arrows are getting shorter as they spread out, the divergence may be zero or negative.

In the exercises you will construct an mfile, `divg.m`, that is the analog of `curl.m`. It superimposes the vector field on a pcolor plot of the $\mathrm{div}(\mathbf{F})$.

Exercises

1. a) Let $\mathbf{F}(x, y) = [xy, \cos(xy)]$. Use `quiver` to graph the vector field on the square $0 \leq x, y \leq 4$.

b) Let C be parameterized by $\mathbf{r}(t) = [t^2, \exp(t)]$, $0 \le t \le 2$. Following Example 11.3, make a numerical estimate of the line integral $\int_C \mathbf{F} \cdot d\mathbf{r}$. Use Simpson's rule, first with 100 subdivisions and then with 200 subdivisions. Estimate the error in the first calculation.

2. a) Let $\mathbf{F}(x, y, z) = [1 - y^2 - z^2, 0, 0]$. Use `quiver3` to graph the vector field in the cube $-1 \le x, y, z \le 1$.

b) Let the curve C be parameterized by $\mathbf{r}(t) = [t, t\cos t, t\sin t]$, $0 \le t \le 2\pi$. Calculate the line integral $\int_C \mathbf{F} \cdot d\mathbf{r}$ by hand, and numerically with Simpson's rule.

3. Let the curve C be given in tabular form,

x	0	.5083	.3833	-.5825	-2.0333	-3.1416
y	0	.3693	1.1951	1.7927	1.4733	-0.0000

x	-3.1416	-3.0499	1.3591	1.5533	4.5749	6.2832
y	-0.0000	-2.2159	-4.1830	-4.7805	-3.3238	0.0000

Let the vector field $\mathbf{F} = [xy, \cos(xy)]$. Estimate the line integral $\int_C \mathbf{F} \cdot d\mathbf{r}$ using the procedure of Example 11.4.

4. Add a vector of z values to the x and y values of Exercise 3, $z(j) = (j - 1)/10$, $j = 1, \ldots, 11$. Now suppose the discrete points (x_j, y_j, z_j), $j = 1, \ldots, 11$, are located on the curve C. Let the vector field $\mathbf{F} = [xz, \cos(xy), z]$. Estimate the line integral $\int_C \mathbf{F} \cdot d\mathbf{r}$.

5. Let the force field $\mathbf{F}(x, y) = [\sin(xy), x - y]$.

a) Let `corners` be $[0\ 2.2\ 0\ 2.2]$ and let C be the path consisting of the two straight-line segments from $(.5, .5)$ to $(1.5, 1)$ and from $(1.5, 1)$ to $(2, 2)$. Use the mfile `lint.m` to calculate the work integral $\int_C \mathbf{F} \cdot d\mathbf{r}$.

b) Repeat the work calculation with the path C now consisting of 10 line segments forming a curve from $(.5, .5)$ to $(2, 2)$.

c) Try various paths to see how you can make the work along the path C from $(.5, .5)$ to $(2, 2)$ nonnegative and as small as possible.

6. a) Use the same vector field over the same rectangle as in Exercise 5. Calculate the work around a closed path with 10 segments using the mfile `lint.m`. First do a closed path in the lower right-hand corner, then in the upper left-hand corner, and finally in the upper right-hand corner. Calculate and record the work/area for each closed path.

b) Use the mfile `curl.m` to see where curl($\mathbf{F}$) is greatest and where smallest. Compare with your results in part a). If in each case you divide the work (circulation) by the area of the enclosed polygon, how close do you come to the value of the curl($\mathbf{F}$) at the center of the polygon?

7. Let $\mathbf{F}(x, y) = [x + \sin y, \; x \cos y]$.

a) Use the mfile `lint.m` with this $\mathbf{F}$ and the rectangle $R = [1, 4] \times [-2, 2]$. Calculate the work done by any two paths from the point $(1.5, -1)$ to the point $(3.5, 1)$. Use any number of segments. What are your results?

b) Calculate the work (circulation) around any closed path. What can you conclude about this vector field?

★   **8.** Let the vector field $\mathbf{F} = [-y, x]/r^p$, where $r = \sqrt{x^2 + y^2}$.

a) Calculate curl($\mathbf{F}$) by hand and show that

$$\mathrm{curl}(\mathbf{F}) = \frac{2 - p}{r^p}\mathbf{k}.$$

This result appears to challenge our intuition, because the flow is counterclockwise for all $p > 0$, yet the curl($\mathbf{F}$) points in the negative $\mathbf{k}$ direction for $p > 2$.

b) Let $p = 3$. Use `lint.m` on the rectangle $R = [.5, 1.5] \times [-.5.5]$. Calculate the circulation in the counterclockwise direction around a small square centered at $(1, 0)$, where curl($\mathbf{F}$) $= -\mathbf{k}$. Note that the work W_1 done on the right side of the square is positive but that the work done on the left side of the square is negative, $W_3 < 0$, and that $W_1 + W_3 < 0$. The work done on the top and bottom segments is of a smaller order.

c) Let $p = 1$. Do the same calculation and note now that $W_1 + W_3 > 0$. Does this agree with the hand calculation done in part a)?

9. Let a vector $\mathbf{p} = [p_1, p_2]$ be given. Let $a = a(s)$. Let $\mathbf{F}(x, y) = a(p_2 x - p_1 y)\mathbf{p}$. This is an example of a shear flow (see Example 11.6).

a) Calculate the curl($\mathbf{F}$) by hand.

b) Let $\mathbf{p} = [1, 2]$ and $a(s) = |s|$. Use `lint.m` on the rectangle $R = [1, 3] \times [1, 3]$. Calculate the circulation around a small closed path lying above the line $y = x/2$ and around a small closed path lying below the line $y = x/2$. Do the results agree with your calculation in part a)?

10. Let $\mathbf{F}(x, y) = [1, (y/4) \sin(\pi x)]$.

a) Calculate div($\mathbf{F}$) by hand. Where is it positive, and where is it negative?

b) Use the mfile `flux2.m` over the rectangle $R = [0, 2] \times [0, 2]$. Calculate the flux around a small square lying in $x < 1$, and divide by the area. Does this approximate div($\mathbf{F}$) at the center of the square?

c) Do the same calculation over a small square lying in $x > 1$, and answer the same question as in part b).

11. The mfile `curl.m` uses centered differences to approximate the value of $v_x - u_y$ in the lines `C1 = ...` and `C2 = ....`

a) Use the centered difference formula

$$\frac{u(x + h, y) - u(x - h, y)}{2h} + \frac{v(x, y + h) - v(x, y - h)}{2h}$$

to approximate the divergence $u_x + v_y$. Modify the mfile `curl.m` to make a pcolor plot of the divergence of a vector field $\mathbf{F}(x, y) = [u(x, y), v(x, y)]$. Call the new file `divg.m`.

b) Use `divg.m` to plot the divergence of the vector field of exercise 10.

12. Let the vector $\mathbf{p} = [p_1, p_2]$ and let $\mathbf{F}(x, y) = a(p_1 x + p_2 y)\mathbf{p}$. This is accelerated, parallel flow (see Example 11.10).

a) Calculate div($\mathbf{F}$) by hand.

b) Let $\mathbf{p} = [2, -1]$ and let $a(s) = |s| + s^2$. Use the mfile `flux2.m` over the rectangle $R = [-2, 2] \times [-2, 2]$. Calculate the flux around a small square above the line $y = 2x$ and divide by the area. Does this approximate the divergence and the center of the square?

c) Repeat the calculation for a small square lying below the line $y = 2x$. Answer the same question.

13. Let $\mathbf{F} = [x, y]/r^p$, with $r = \sqrt{x^2 + y^2}$.

a) Calculate div($\mathbf{F}$) by hand. Show that

$$\text{div}(\mathbf{F}) = \frac{2 - p}{r^p}.$$

Again, this result seems counterintuitive. Although the arrows are spreading apart for all $p > 0$, the divergence is negative for $p > 3$. This can be understood by noting that the larger p is, the faster the arrows are getting shorter as r increases.

b) Let $p = 3$. Use the mfile `flux2.m` on the rectangle $R = [.5, 1.5] \times [-.5, .5]$. Calculate the flux around a square with side .4 centered at $(1, 0)$. Note that the flux along the top and bottom of the square f_1 and f_3 are both positive, with $f_1 \approx f_3$. The flux f_2 on the right side is positive, while the flux f_4 on the left side is negative with $f_1 + f_2 + f_3 + f_4 < 0$.

c) Now let $p = 1$. Repeat the calculation of part b). What is the sign of the flux? How do the individual parts add up?

d) Finally, let $p = 2$. Repeat the calculation of part b). What is the result? How do the individual parts balance out?

★★   **14.** Let $\mathbf{F}(x, y) = [f(x, y), g(x, y)]$ be a vector field and let G be the unit square $0 \le x, y \le 1$. The flux through the boundary of G (using the exterior normal) is the sum of four integrals:

$$\int_{\partial G} \mathbf{F} \cdot \mathbf{n} \, ds = -\int_0^1 f(0, y)dy + \int_0^1 g(x, 1)dx + \int_0^1 f(1, y)dy - \int_0^1 g(x, 0)dx.$$

a) Write a script file to calculate this flux integral. The script should call function mfiles f.m and g.m. Use the one-dimensional Simpson's rule with $n = 20$ to calculate each of the four integrals.

b) To test your script, use it to calculate the flux integral for $\mathbf{F}(x, y) = [(x + 1)y, \; 3y^2]$, and compare with your calculation by hand.

c) Now let $f(x, y) = \exp(x + y)$ and $g(x, y) = y \cos(x^2)$. Calculate the flux integral with your script. Then use the mfile simp2 to calculate numerically the integral $\int \int_G \text{div}(\mathbf{F}) \, dxdy$ with $n = m = 20$ and then with $n = m = 40$. Compare the result for the double integral with the result for the boundary integrals.

12

Problems from Electrostatics and Fluid Flow

Prepared mfiles used in this chapter

```
flow1   flow2
```

Chapter 12 is divided into two major sections, one dealing with problems from electrostatics and the other covering problems from fluid flow. We shall discuss electrostatics first because we deal with a single equation. In the discussion of fluid flow, we shall need to consider systems of equations.

First we introduce some techniques we will use for deriving the equations.

12.1   An important tool

The principle tool used to derive the partial differential equations of physics is the divergence theorem. The divergence theorem connects global observations of vector fields (e.g., flux integrals over surfaces) with the local behavior of the vector fields (partial derivatives of the components). Application of the divergence theorem often leads to an integral equation, for instance,

$$\int\int_{G} f(x, y)\, dA(x, y) = 0 \qquad \text{or} \qquad \int\int\int_{G} f(x, y, z)\, dV(x, y, z) = 0,$$

(12.1)

holding for all sets G in xy space or all sets G in xyz space. Next we want to deduce a statement about the point values of f from (12.1).

241

Null Theorem

If f is continuous and Eq. (12.1) holds for all sets G, then $f \equiv 0$.

We show why this is true in two dimensions. The argument in three dimensions is the same. Fix a point (x_0, y_0). Condition (12.1) holds in particular for all squares Q_h of side h centered at (x_0, y_0),

$$Q_h = \{(x, y) : |x - x_0| \leq h/2, \ |y - y_0| \leq h/2\}.$$

Dividing Eq. (12.1) by h^2, we have

$$\frac{1}{h^2} \int \int_{Q_h} f(x, y) \, dxdy = 0 \qquad \text{for all } h > 0. \qquad (12.2)$$

But Eq. (12.2) is just the average of f over Q_h. As $h \downarrow 0$, this average, which is always zero by Eq. (12.2), tends to the value of f at (x_0, y_0). We write this as

$$f(x_0, y_0) = \lim_{h \to 0} \frac{1}{h^2} \int \int_{Q_h} f(x, y) \, dxdy = 0.$$

Since (x_0, y_0) could be any point, we see that f is zero everywhere.

12.2 Electrostatics

In this introduction to the subject, we shall set all the physical constants to 1. The electric field $\mathbf{E} = [E_1, E_2, E_3]$ is a vector attached to each point in some set G of xyz space. It is the force exerted on a particle with a unit positive charge. The field is produced by the presence of charge, according to Coulomb's law. If a charge of Q coulombs is located at the point $\mathbf{x_0} = (x_0, y_0, z_0)$, the electric field at the point $\mathbf{x} = (x, y, z)$ is

$$\mathbf{E}(x, y, z) = \frac{Q(\mathbf{x} - \mathbf{x_0})}{4\pi \|\mathbf{x} - \mathbf{x_0}\|^3}. \qquad (12.3)$$

Notice that if $Q > 0$, the electric field vector $\mathbf{E}$ points from $\mathbf{x_0}$ to $\mathbf{x}$. In this case, the force is repellent. If $Q < 0$, the $\mathbf{E}$ vector points from $\mathbf{x}$ to $\mathbf{x_0}$, and the charge at $\mathbf{x_0}$ is attracting the positive unit test charge. In either case, the magnitude of $\mathbf{E}$ is

$$\|\mathbf{E}(x, y, z)\| = \frac{|Q|}{4\pi \|\mathbf{x} - \mathbf{x_0}\|^2}.$$

This is an example of an *inverse square law*.

What is the electric field produced by a distribution of charge? Two facts about the electric field generated by such a charge distribution, which can be verified in the laboratory, are

I   The electric field is conservative. The line integral $\int_C \mathbf{E} \cdot d\mathbf{r} = 0$ for all closed paths C.

II   Gauss's law The strength of the electric field, as measured by the flux integral

$$\int\int_S \mathbf{E} \cdot \mathbf{n} \, dS$$

over a closed surface S, is equal to the total charge contained within the surface. When the charge is distributed continuously over a volume by a charge density $\rho(x, y, z)$, Gauss's law becomes

$$\int\int_S \mathbf{E} \cdot \mathbf{n} \, dS = \int\int\int_G \rho(x, y, z) \, dV, \qquad (12.4)$$

where G is the region bounded by S. These two facts will allow us to determine the electric field from a given charge density ρ.

Assuming our region of investigation has no holes in it, the fact that $\mathbf{E}$ is conservative means that there is a potential function $\phi(x, y, z)$ such that

$$-\nabla\phi = \mathbf{E}. \qquad (12.5)$$

The minus sign is a convention used in physics. A surface on which ϕ is constant is called an *equipotential surface*.

Now we apply the divergence theorem to the integral on the left of Eq. (12.4). We deduce that

$$\int\int\int_G \mathrm{div}(\mathbf{E}) \, dV = \int\int\int_G \rho \, dV \qquad (12.6)$$

or

$$\int\int\int_G [\mathrm{div}(\mathbf{E}) - \rho] \, dV = 0$$

for all bounded sets G. Hence by the Null theorem, we can conclude that

$$\mathrm{div}(\mathbf{E})(x, y, z) = \rho(x, y, z). \qquad (12.7)$$

But from Eq. (12.5), we have $\mathrm{div}(\mathbf{E}) = \mathrm{div}(-\nabla\phi) = -\Delta\phi$. Thus the potential ϕ must satisfy

$$-\Delta\phi = \rho. \qquad (12.8)$$

This is the *Poisson equation*. It appears in many other contexts, including heat transfer and acoustics. The operator on the left, $\Delta\phi = \phi_{xx} + \phi_{yy} + \phi_{zz}$, is the Laplace operator seen in Chapter 8.

Now we have a single partial differential equation that relates the electrostatic potential to a given charge density. Once ϕ is determined, $\mathbf{E}$ is found from Eq. (12.5).

The physically relevant solution of Eq. (12.8), in the absence of any boundaries, is found using the potential of a point charge. It is easy to verify that the potential of the electric field (Eq. 12.3) for $Q = 1$ and $\mathbf{x}_0 = 0$ is

$$\gamma(\mathbf{x}) = \frac{1}{4\pi \|\mathbf{x}\|}.$$

γ is the fundamental potential in three dimensions. You should verify that $\Delta\gamma = 0$ for $\mathbf{x} \neq 0$. If the unit charge is placed at $\mathbf{y} \neq 0$, the potential is

$$\gamma(\mathbf{x} - \mathbf{y}) = \frac{1}{4\pi \|\mathbf{x} - \mathbf{y}\|}.$$

Notice that the potential γ becomes infinite as $\mathbf{x} \to \mathbf{y}$ because the unit charge is concentrated at a single point $\mathbf{y}$.

The potential produced by charges Q_1 at $\mathbf{y}_1$ and Q_2 at $\mathbf{y}_2$ is

$$Q_1\gamma(\mathbf{x} - \mathbf{y}_1) + Q_2\gamma(\mathbf{x} - \mathbf{y}_2).$$

If charges $Q_1, Q_2, \ldots, Q_n$ are placed at locations $\mathbf{y}_1, \mathbf{y}_2, \ldots, \mathbf{y}_n$, the resulting potential is

$$\phi(\mathbf{x}) = \sum_{j=1}^{n} Q_j\gamma(\mathbf{x} - \mathbf{y}_j) = \sum_{j=1}^{n} \frac{Q_j}{4\pi \|\mathbf{x} - \mathbf{y}_j\|}. \tag{12.9}$$

For a continuous distribution of charge, given by a density $\rho(\mathbf{x})$, the discrete sum (Eq. 12.9) is replaced by

$$\phi(\mathbf{x}) = \int\int\int \gamma(\mathbf{x} - \mathbf{y})\rho(\mathbf{y}) \, dV(\mathbf{y}) \tag{12.10}$$

$$= \frac{1}{4\pi} \int\int\int \frac{\rho(\mathbf{y})}{\|\mathbf{x} - \mathbf{y}\|} \, dV(\mathbf{y}).$$

A Riemann sum approximation to this integral is

$$\frac{1}{4\pi} \sum_{i,j,k} \frac{\rho(\mathbf{y}_{i,j,k})}{\|\mathbf{x} - \mathbf{y}_{i,j,k}\|} \Delta V, \tag{12.11}$$

where we have decomposed the region of integration into cubes $R_{i,j,k}$ of volume ΔV and picked a point $y_{i,j,k} \in R_{i,j,k}$. In the exercises you will write a MATLAB code that computes this approximation. This is not the approach used in efficient, accurate, computation of solutions to this problem, but it will give us some idea of what the solutions look like.

Example 12.1

We compute and graph the potential $\phi(x, y, z)$ produced by a uniform charge density $\rho \equiv 1$ on the solid cylinder $R = \{x^2 + y^2 < 1, -1 \leq z \leq 1\}$. It is given by the triple integral, which can be put in cylindrical coordinates:

$$\phi(x, y, z) = \frac{1}{4\pi} \int \int \int_R \frac{dV(y)}{\|x - y\|}$$

$$= \frac{1}{4\pi} \int_{-1}^{1} \int_0^1 \int_0^{2\pi} \frac{r \, dr \, d\theta \, d\zeta}{\sqrt{(x - r\cos\theta)^2 + (y - r\sin\theta)^2 + (z - \zeta)^2}}$$

Since the potential is symmetric with respect to the z axis, it suffices to compute values of the potential $\phi(x, 0, z)$ and display the results in the x, z plane. The integral to compute is

$$\phi(x, 0, z) = \frac{1}{4\pi} \int_{-1}^{1} d\zeta \int_0^1 r \, dr \int_0^{2\pi} \frac{d\theta}{\sqrt{x^2 - 2xr\cos\theta + r^2 + (z - \zeta)^2}}.$$

We do this in the following script. The graph of $\phi(x, 0, z)$ is displayed in the left side of Figure 12.1, and the level curves of $\phi(x, 0, z)$ are displayed on the right. The equipotential surfaces of ϕ are generated by revolving these curves about the z axis.

```
% choose midpoints in the r, theta, zeta grid.
delr = .1; deltheta = pi/10; delzeta = .1;
r = linspace(.5*delr, 1-.5*delr, 10);
th = linspace(.5*deltheta, 2*pi-.5*deltheta, 20);
zeta = linspace(-1+.5*delzeta, 1-.5*deltheta, 20);
% We form the three-dimensional arrays, which are
%   10 by 20 by 20
[R,TH,ZETA] = meshgrid(r,th,zeta);

% construct the meshgrid for graphing the potential.
x = linspace(0, 5, 26);
z = linspace(-5,5, 51);
[X,Z] = meshgrid(x,z);
```

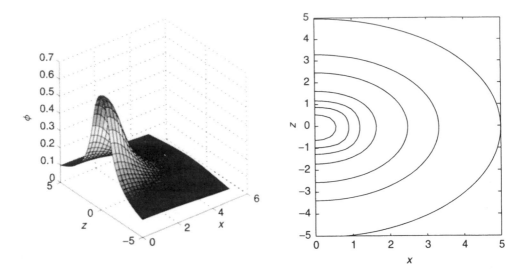

Figure 12.1 Potential ϕ produced by uniform charge density on the cylinder $R = \{x^2 + y^2 < 1, -1 \le z \le 1\}$. Graph of $\phi(x, 0, z)$ on the left, level curves on the right.

```
% set aside a 51 by 26 array for the values of phi.
phi = 0*Z;

for i = 1:51
    for j = 1:26
        % for each z(i),x(j) in the x,z meshgrid,
        % compute the three-dimensional array of the denominator
        denom =
        sqrt(x(j)^2-2*x(j)*R.*cos(TH)+R.^2+(z(i)-ZETA).^2);
        % compute the Riemann sum by summing over the three-
        % dimensional arrays R and denom
        phi(i,j) =
        (delr*deltheta*delzeta/(4*pi))*sum(sum(sum(R./denom)));
    end
end
subplot(1,2,1)
    surf(X,Z,phi)
subplot(1,2,2)
    contour(X,Z,phi)
```

The Dirichlet problem

In a region G with no charge, the equation for the potential is simply $\Delta\phi = 0$. Solutions of this equation are called *harmonic functions* (see Chapter 8), and they have very special properties. This situation arises when G is the interior of a shell bearing charge. The charge on the shell produces an electric field and hence a potential ϕ. The values of ϕ inside of G are determined by its values on the bounding shell, which we shall denote by ∂G. The *Dirichlet problem* is to find the solution of the boundary value problem

$$\Delta\phi(\mathbf{x}) = 0 \quad \text{in } G, \qquad \phi = f \quad \text{on } \partial G. \tag{12.12}$$

The function f is given on ∂G. These are the values the potential is supposed to have on ∂G.

In more advanced courses it is shown that if the region has no sharp cusps and f is continuous, then there is a unique solution to the Dirichlet problem. Furthermore the solution can be represented in terms of f, much as in Eq. (12.10). There is a function $P(\mathbf{x}, \mathbf{y})$, determined by the domain G, such that the solution of Eq. (12.12) is given by

$$\phi(\mathbf{x}) = \int\int_{\partial G} P(\mathbf{x}, \mathbf{y}) f(\mathbf{y}) \, dS(\mathbf{y}). \tag{12.13}$$

$P(\mathbf{x}, \mathbf{y})$ is called the *Poisson kernel* for the region G. For $\mathbf{y} \in \partial G$, $\mathbf{x} \to P(\mathbf{x}, \mathbf{y})$ satisfies the Laplace equation inside G.

For $\mathbf{y} \in \partial G$, as $\mathbf{x} \to \partial G$, $P(\mathbf{x}, \mathbf{y}) \to 0$, except for the special case when $\mathbf{x} \to \mathbf{y}$ when P blows up.

The Poisson kernel can be thought of as the potential produced by a unit charge placed at the point $\mathbf{y} \in \partial G$ with the boundary grounded (potential equals 0) everywhere except at $\mathbf{y}$.

The Poisson kernel can be found in closed form only for special geometries, such as a sphere, circle, half-plane, or quarter-plane. For this reason, Eq. (12.13) is not used for computation of solutions. Nevertheless, it can give us some insight into the qualitative properties of the solution and how it depends on the boundary data. In the exercises we shall examine the Poisson kernel for the circle and the half-plane.

12.3   The geometry of fluid flow

By a *fluid* we mean either a gas or a liquid. We shall limit our treatment to fluid flow in two dimensions, but the derivation of the equations is easily extended to three dimensions. Furthermore we shall limit our discussion to the case of steady flow. In this situation the velocity is independent of time. This does not mean the fluid particles are stationary; they can move.

We shall first investigate how certain quantities move with the fluid for a given fluid velocity field. In the next section, we apply the laws of physics to find which velocity vector fields can represent a fluid.

In the Eulerian description of a fluid, the observer views the fluid at fixed points (x, y) in space. Her observation point does not move with the fluid. At each point in the two-dimensional space considered, we assume given the velocity of the fluid

$$\mathbf{q} = [u(x, y), v(x, y)].$$

The x component of the velocity is u, and the y component is v. Now, from our observer's point of view, if we wish to follow the path of a fluid particle, we need functions $x(t)$ and $y(t)$ to locate the particle as a function of t. The curve $t \to (x(t), y(t))$ parameterizes the path of the particle, and the velocity of this particle at time t is

$$[x'(t), y'(t)].$$

This expression must be the same as the given velocity vector at the location $(x(t), y(t))$. Consequently, $x(t)$ and $y(t)$ must satisfy the following system of differential equations:

$$\frac{dx}{dt} = u(x(t), y(t)) \qquad (12.14)$$

$$\frac{dy}{dt} = v(x(t), y(t)).$$

For a given velocity field $[u, v]$, Eqs. (12.14) can be very difficult or impossible to solve analytically. Instead we shall use a numerical method to generate approximate solutions to Eqs. (12.14). The method we describe is simple but effective in certain situations. It is called the *Euler* method, and it is the starting point for the construction of more sophisticated methods.

We suppose that at time $t = 0$, the particle is at the point (x_1, y_1). The velocity vector at that point is

$$\mathbf{q}_1 = [u(x_1, y_1), v(x_1, y_1)].$$

Over a very short time interval Δt, the particle will move from the point (x_1, y_1) to approximately

$$(x_2, y_2) = (x_1 + \Delta t u(x_1, y_1), \ y_1 + \Delta t v(x_1, y_1)).$$

We use this approximation again, starting from (x_2, y_2), which yields

$$(x_3, y_3) = (x_2 + \Delta t u(x_2, y_2), \ y_2 + \Delta t v(x_2, y_2)).$$

We can repeat this procedure any number of times. The general form is

$$(x_{n+1}, y_{n+1}) = (x_n + \Delta t u(x_n, y_n), \; y_n + \Delta t v(x_n, y_n)). \qquad (12.15)$$

Example 12.2

The following script implements the Euler method for the velocity field

$$u(x, y) = x + y, \qquad v(x, y) = \cos(y).$$

We assume u and v are given as inline functions or in mfiles, u.m and v.m.

```
start = input('enter the starting point [x1,y1]')
delt = input('enter the time step delta t   ')
nstep = input('enter the number of time steps ')
x = zeros(1,nstep+1); y = x;
x(1) = start(1);
y(1) = start(2);

for n = 1:nstep
    x(n+1) = x(n) + delt*u(x(n),y(n));
    y(n+1) = y(n) + delt*v(x(n),y(n));
end
plot(x,y,'r')
```

You can plot this polygonal path together with the velocity field $[u, v]$ using a meshgrid and the command quiver (see Figure 12.2). You can further compare the path produced by the Euler method with that produced by the mfile flow1.m, which uses a more sophisticated numerical solver of the system (12.14).

mfile flow1

The mfile flow1.m solves the system (12.14) numerically using the more sophisticated solver ode45 of MATLAB. It also plots the vector field. The call is flow1(corners,T), where, as usual, corners is the vector of corner coordinates $[a, b, c, d]$ of the rectangle where the vector field is to be displayed. T is the time up to which the flow is followed. The vector field components must be provided in mfiles u.m, and v.m. flow1.m also requires another mfile wdot.m, which is provided. wdot.m uses the mfiles u.m and v.m.

After the call, the program asks the user to enter the number M of starting points for the trajectories. After M is entered, the program waits for the user to click on the figure for a starting point. The program then computes the trajectory through this point and superimposes it on the vector field. This can be done M times. If the trajectory goes outside the rectangle determined by corners, the time T should be shortened.

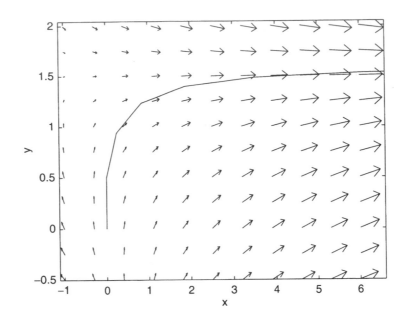

Figure 12.2 Vector field $\mathbf{q} = [u, v] = [x + y, \cos(y)]$. Approximate particle path starting at $(0, 0)$ computed by Euler method with $\Delta t = .5$, 10 time steps.

Example 12.3

Let the vector field be $u(x, y) = -y + x^2$ and $v = x + x^3$. We choose the rectangle $R = [-2.5, 2.5] \times [-1, 6]$. The results of using flow1 with $M = 5$ trajectories is shown in Figure 12.3. The 5 trajectories begin at points $(0, 1)$, $(0, 2)$, $(0, 3)$, $(0, 4)$, $(0, 5)$.

```
function z = u(x,y)
      z = -y +x.^2;

function z = v(x,y)
      z = x + x.^3;

>> corners = [-2.5 2.5 -1 6]
>> flow1(corners, 4)
```

Following the flow

Next we shall examine how a scalar quantity, such as the density of the fluid, changes with the flow. We shall use this result in the next section to find the equations of motion of the fluid.

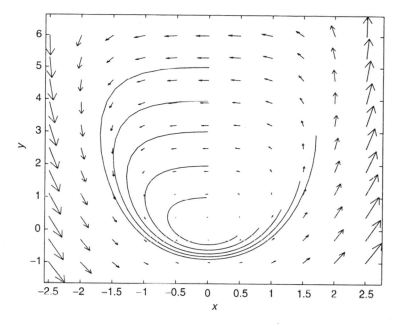

Figure 12.3 Particle paths of the flow $\mathbf{q} = [u, v] = [-y + x^2, x + x^3]$. The flow rotates in the counterclockwise direction.

Let $G = G(0)$ be a bounded set in x, y space consisting of fluid particles. As these particles move following the solution curves of (12.14), the set $G(0)$ is deformed into a set $G(t)$ at time t. The shape and area of $G(t)$ may change with t.

Let $f(x, y)$ be a continuously differentiable function. How does the integral

$$\int \int_{G(t)} f(x, y) \, dx dy \qquad (12.16)$$

depend on t? We give an intuitive derivation of the formula for

$$\frac{d}{dt} \int \int_{G(t)} f(x, y) \, dx dy,$$

assuming $f(x, y) \geq 0$.

Let $\Delta(h)$ be the region between the boundaries of $G(t)$ and $G(t + h)$ (see Figure 12.4). Let $\mathbf{n}$ be the unit exterior normal to $\partial G(t)$. The width of $\Delta(h)$, measured in the direction along the normal $\mathbf{n}$, varies as we move along $\partial G(t)$. At each point of $\partial G(t)$, it is given approximately by $|\mathbf{n} \cdot \mathbf{q}| h$. Where the fluid particles move outward from $G(t)$, i.e., where $\mathbf{n} \cdot \mathbf{q} > 0$, the integral (12.16) will increase.

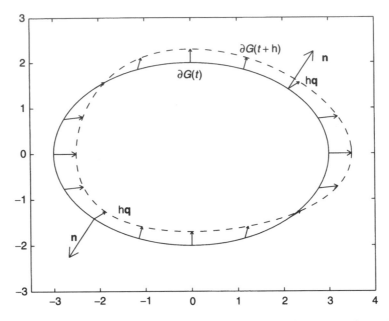

Figure 12.4 Set $G(t)$ displaced a short time interval $h > 0$ by a flow. Boundary of $G(t)$ in solid line, boundary of $G(t + h)$ in dotted line.

Where the fluid particles move into $G(t)$, i.e., where $\mathbf{n} \cdot \mathbf{q} < 0$, the integral (12.16) decreases. Hence the net change

$$\int\int_{G(t+h)} f(x, y)\, dxdy - \int\int_{G(t)} f(x, y)\, dxdy \approx \int\int_{\Delta(h)} f(x, y)\mathrm{sign}(\mathbf{n} \cdot \mathbf{q})\, dxdy$$

$$\approx h \int_{\partial G(t)} f\, \mathbf{n} \cdot \mathbf{q}\, ds.$$

Dividing by h and taking the limit as $h \to 0$, we have

$$\frac{d}{dt} \int\int_{G(t)} f(x, y)\, dxdy = \int_{\partial G(t)} f\, \mathbf{n} \cdot \mathbf{q}\, ds. \qquad (12.17)$$

Now finally, we use the divergence theorem in a crucial way to convert the boundary integral on the right into a double integral over $G(t)$. The vector field in this case is the product of f times the velocity vector $\mathbf{q}$. Hence the integral on the right of Eq. (12.17) can be written

$$\int_{\partial G(t)} f\mathbf{n} \cdot \mathbf{q}\, ds = \int\int_{G(t)} \mathrm{div}(f\mathbf{q})\, dxdy,$$

which yields the fundamental result

$$\frac{d}{dt} \int \int_{G(t)} f(x, y) \, dxdy = \int \int_{G(t)} \text{div}(f\mathbf{q}) \, dxdy. \qquad (12.18)$$

We shall make several choices of f to derive the equations of the fluid. However, to get an immediate, geometrically important, result, we take $f \equiv 1$. Then

$$\int \int_{G(t)} f(x, y) \, dxdy = A(t),$$

the area of $G(t)$. Putting $f \equiv 1$ in the right side of Eq. (12.18) yields

$$\frac{d}{dt} A(t) = \int \int_{G(t)} \text{div}(\mathbf{q}) \, dxdy. \qquad (12.19)$$

Thus whether the area of $G(t)$ decreases or grows with the flow depends on the sign of div($\mathbf{q}$). If div($\mathbf{q}$) > 0, $G(t)$ expands as t increases; if div($\mathbf{q}$) < 0, $G(t)$ shrinks as t increases. In the special case that div($\mathbf{q}$) $= 0$ (incompressible flow), the area of $G(t)$ is constant, although the shape may change.

mfile flow2

The mfile flow2.m works like flow1.m and allows the user to follow the deformations of a disk carried with the flow. The call is flow2(corners,times). As usual, corners is the vector $[a, b, c, d]$ of corner coordinates of the rectangle R where the vector field is displayed. times = [t1 t2 t3 t4] is a vector of the four times at which the deformed disk is plotted, in addition to $t = 0$. The vector field components, $u(x, y)$ and $v(x, y)$, must be provided in mfiles u.m and v.m. The mfile wdot.m is also needed. After the call, the vector field is displayed in the rectangle R, and the program waits for the user to click on the figure to determine the center of the initial disk. The area of the deformed disks is computed approximately, using a numerical approximation to the line integral $\int_C x\,dy$.

Example 12.4

Let the flow be given by the vector field

$$\mathbf{q} = [u, v] = [x + y, \cos(y)].$$

The initial disk is centered at $(-.2, -.5)$, and the times are $t_1 = .4, t_2 = .8, t_3 = 1.2, t_4 = 1.6$. The areas of the initial disk and the images are

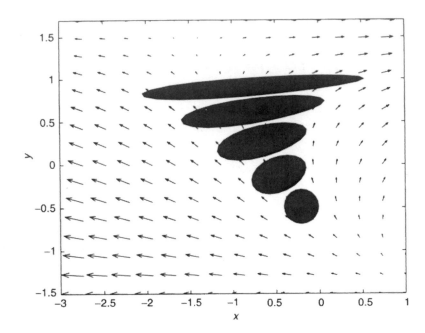

Figure 12.5 Images of a disk centered at $(-.2, -.5)$ carried by the flow $[u(x, y), v(x, y)] = [x + y, \cos y]$ at times $t_1 = .4, t_2 = .8, t_3 = 1.2, t_4 = 1.6$.

```
area0  =  .1236
area1  =  .2078
area2  =  .3005
area4  =  .4262
```

See Figure 12.5. From the figure and the values of the areas, it is clear that $\mathrm{div}(\mathbf{q}) > 0$ in this region. In fact, $\mathrm{div}(\mathbf{q}) = 1 - \sin y > 0$ for $-3\pi/2 < y < \pi/2$.

12.4   The Euler equations

Up to now we have described the geometry of the motion of a fluid. We have not invoked any physical principles that govern the flow of a fluid. We do this now to derive the equations of motion. Remember that we are assuming that the velocity $\mathbf{q} = [u(x, y), v(x, y)]$ does not depend on t.

Two additional quantities that we will need to describe a fluid are the pressure $p(x, y)$ and the density $\rho(x, y)$. Since we are working in two dimensions, the units of ρ are mass/area; for any region G of the fluid, $\int\int_G \rho\, dx dy$ is the mass of the fluid particles contained in G. If we have a region $G(t)$ that moves with the flow, always consisting of the same fluid particles, the mass of fluid contained in $G(t)$ is constant. Hence

$$\frac{d}{dt} \int \int_{G(t)} \rho(x, y) \, dxdy = 0. \qquad (12.20)$$

Now we apply Eq. (12.18), taking $f(x, y) = \rho(x, y)$. We conclude that

$$0 = \int \int_{G(t)} \text{div}(\rho\mathbf{q}) \, dxdy \qquad (12.21)$$

for any set $G(t)$. By the Null theorem, assuming ρ and $\mathbf{q}$ are continuously differentiable, we see that ρ and $\mathbf{q}$ must satisfy the partial differential equation

$$\text{div}(\rho\mathbf{q}) = \nabla\rho \cdot \mathbf{q} + \rho \, \text{div}(\mathbf{q}) = 0. \qquad (12.22)$$

This is our first equation of motion. It expresses the physical law of *conservation of mass*.

Next we turn our attention to the momentum of the fluid. The momentum density is a vector field $\rho\mathbf{q} - [\rho u(x, y), \rho v(x, u)]$. We shall assume that the fluid has no viscosity, and that the only force acting on a body of fluid is the pressure. Newton's second law states that the rate of change of momentum is equal to the applied force (in this case the pressure). We follow the same procedure we did to derive the conservation of mass equation, and we invoke Newton's second law on each component of the momentum density. This yields equations for the x and y components of the momentum:

$$\nabla(\rho u) \cdot \mathbf{q} + \rho u \, \text{div}(\mathbf{q}) + p_x = 0 \qquad (12.23)$$

and

$$\nabla(\rho v) \cdot \mathbf{q} + \rho v \, \text{div}(\mathbf{q}) + p_y = 0. \qquad (12.24)$$

These equations express *conservation of momentum*.

These equations may be simplified by taking into account Eq. (12.22). Since $\nabla(\rho u) = u\nabla\rho + \rho\nabla u$, we have

$$\begin{aligned}
\nabla(\rho u) \cdot \mathbf{q} + \rho u \, \text{div}(\mathbf{q}) &= u\nabla\rho \cdot \mathbf{q} + \rho\nabla u \cdot \mathbf{q} + \rho u \, \text{div}(\mathbf{q}) \\
&= u \, \text{div}(\rho\mathbf{q}) + \rho\nabla u \cdot \mathbf{q} \\
&= \rho\nabla u \cdot \mathbf{q}.
\end{aligned}$$

The same holds for the left side of Eq. (12.24). Hence our equations for steady motion of a fluid are

$$\text{div}(\rho \mathbf{q}) = 0 \tag{12.25}$$

$$\rho \mathbf{q} \cdot \nabla u = -p_x \tag{12.26}$$

$$\rho \mathbf{q} \cdot \nabla v = -p_y. \tag{12.27}$$

These are known as the *Euler equations* of steady flow. The second and third equations are often written as a single vector equation,

$$\mathbf{q} \cdot \nabla \mathbf{q} = -\nabla p / \rho.$$

We note that the Euler equations are a system of three partial differential equations in the four unknown functions ρ, u, v, and p. Generally, for a system to be solvable there must the same number of equations as there are unknown functions. For this reason, we add a fourth equation to the system (12.25)–(12.27) that relates the pressure and the density. This is called the *equation of state*. In the absence of any temperature variation, or with very little temperature variation, the equation of state is written

$$F(\rho, p) = 0. \tag{12.28}$$

The equation of state is determined by the type of fluid. For example, in a gas, such as air, the equation of state is

$$p = A\rho^\gamma,$$

where A and γ are constants, $\gamma > 1$.

A *solution* of Eqs. (12.25)–(12.28) is a set of four functions $u(x, y)$, $v(x, y)$ $\rho(x, y)$, and $p(x, y)$ that satisfy the equations everywhere in a specified region of the xy plane. It is very difficult to find solutions of these equations in closed form. More often, approximate numerical solutions are found using high-speed computers

A fluid such as water or oil is practically incompressible, so the density is nearly constant, although the pressure may vary. When the density is constant, Eq. (12.25) becomes

$$u_x + v_y = \text{div}(\mathbf{q}) = 0, \tag{12.29}$$

which is the *incompressiblity condition*. In this case, the equation of state is no necessary.

12.5   Incompressible flow

Most solutions in closed form are for incompressible flow, where ρ is constant. The incompressibility condition (12.29) may be rewritten

$$0 = u_x + v_y = u_x - (-v_y)$$

so that the vector field $[-v, u]$ is the gradient of a scalar function $\psi(x, y)$:

$$\psi_x = -v, \qquad \psi_y = u.$$

ψ is called the *stream function*. It is easy to verify that ψ is constant on the particle trajectories, which are now called *streamlines*. This follows from the fact that $\nabla \psi \cdot \mathbf{q} = [-v, u] \cdot [u, v] = -uv + uv = 0$.

Examples 12.5

In these examples of incompressible flow, the density ρ is constant, so we may as well set $\rho \equiv 1$. Our problem comes down to finding velocity fields $\mathbf{q} = [u, v]$ such that $\mathrm{div}(\mathbf{q}) = 0$ and Eqs. (12.26) and (12.27) are satisfied. This will be the case if $\mathbf{q} \cdot \nabla \mathbf{q}$ is a gradient, which is to say $\mathrm{curl}(\mathbf{q} \cdot \nabla \mathbf{q}) = 0$. Then we may take for the pressure p any function such that $-\nabla p = \mathbf{q} \cdot \nabla \mathbf{q}$.

Rotating flow

With $r = \sqrt{x^2 + y^2}$, let $\mathbf{q} = [-y, x]/r^\alpha$. It is easy to verify that $\mathrm{div}(\mathbf{q}) = 0$, and that

$$\mathbf{q} \cdot \nabla \mathbf{q} = -[x, y]/r^{2\alpha},$$

which is a gradient.

For $\alpha \neq 1$, we take the pressure as

$$p(x, y) = \left(\frac{1}{2\alpha - 2}\right) \frac{1}{r^{2\alpha - 2}} + C.$$

For $\alpha = 1$, the pressure is

$$p(x, y) = -\ln r + C.$$

Radial flow

We look for $\mathbf{q}$ in the form $\mathbf{q} = [x, y]/r^\alpha$ and find that $\mathrm{div}(\mathbf{q}) = 0$ only for $\alpha = 2$. In this case $\mathbf{q} \cdot \nabla \mathbf{q} = -[x, y]/r^4$ and

$$p(x, y) = \frac{1}{r^2} + C.$$

It is important to find quantities that are preserved by the flow. This means that we seek scalar quantities that are constant on the streamlines. Of course, the stream function is constant on the streamlines. Another important quantity is a combination of the speed of the flow and the pressure.

Bernoulli's law

Let $\mathbf{q} = [u, v]$, p and $\rho = $ constant be a solution of the Euler Eqs. (12.25)–(12.27). Then the quantity

$$e \equiv \frac{\rho}{2}\|\mathbf{q}\|^2 + p$$

is constant along the streamlines, although it may vary from one streamline to another. To see this, we consider the function of t that is e restricted to a solution curve of Eqs. (12.14),

$$e(t) = \frac{\rho}{2}\|\mathbf{q}(x(t), y(t))\|^2 + p(x(t), y(t)).$$

To show that $e(t)$ is constant, we calculate de/dt. Now,

$$\|\mathbf{q}\|^2 = u^2(x(t), y(t)) + v^2(x(t), y(t)),$$

so by the chain rule,

$$\frac{d}{dt}\frac{\rho}{2}\|\mathbf{q}\|^2 = \rho[uu_x(dx/dt) + uu_y(dy/dt) + vv_x(dx/dt) + vv_y(dy/dt)]$$

$$= \rho[u^2 u_x + uvu_y + uvv_x + v^2 v_y]$$

$$= \rho u(uu_x + vu_y) + \rho v(uv_x + vv_y).$$

Similarly,

$$\frac{d}{dt}p(x(t), y(t)) = p_x(dx/dt) + p_y(dy/dt)$$

$$= p_x u + p_y v.$$

Hence

$$\frac{de}{dt} = \frac{1}{2}\frac{d}{dt}(\rho\|\mathbf{q}\|^2) + \frac{d}{dt}p$$

$$= \rho(uu_x + vu_y)u + \rho(uv_x + vv_y)v + p_x u + p_y v$$

$$= [\rho(uu_x + vu_y) + p_x]u + [\rho(uv_x + vv_y) + p_y]v$$

$$= 0$$

because of Eqs. (12.26) and (12.27).

According to Bernoulli's law, if the speed $\|\mathbf{q}\|$ of the flow increases along a given streamline, then the pressure p must decrease. For example, when a fluid leaves a nozzle, the pressure decreases and the fluid particles accelerate. As we shall see later when we consider irrotational flow, Bernoulli's law is also important in aerodynamics.

Another quantity that is constant along streamlines is the *vorticity*. The vorticity is just the $\mathbf{k}$ component of the curl of $\mathbf{q}$,

$$\omega = v_x - u_y. \tag{12.30}$$

It is a bit more complicated computation to show that $d\omega/dt = 0$ along a streamline, but the method is the same. This result holds only in two dimensions. In three dimensions, the vorticity may change along the streamlines, and this makes the analytical and numerical study of the Euler equations in three dimensions more difficult.

Irrotational, incompressible flow

Finally we make an additional restriction on the type of flows we consider. In addition to assuming that the density ρ is constant, we shall also assume that the flow is *irrotational*,

$$\mathrm{curl}(\mathbf{q}) = 0. \tag{12.31}$$

This means that in any region having no holes, there is a potential function $\phi(x, y)$, called the *velocity potential*, with

$$\nabla\phi = [u, v] = \mathbf{q}.$$

If we now apply the incompressibility condition, $\mathrm{div}(\mathbf{q}) = 0$, we find that ϕ must satisfy the Laplace equation

$$\Delta\phi = \mathrm{div}(\nabla\phi) = \mathrm{div}(\mathbf{q}) = 0. \tag{12.32}$$

This is the same equation that the electrostatic potential satisfies in a region where there is no charge. The stream function ψ also satisfies $\Delta\psi = 0$.

Bernoulli's law becomes even simpler for irrotational, incompressible flow. If ϕ is the velocity potential, then the second and third Euler equations, Eqs. (12.26) and (12.27), are satisfied automatically when we take the pressure to be

$$p(x, y) = -\frac{\rho}{2}\|\mathbf{q}\|^2 + C.$$

In fact,

$$uu_x + vu_y = \phi_x\phi_{xx} + \phi_y\phi_{xy} = \frac{1}{2}\frac{\partial}{\partial x}(\phi_x^2 + \phi_y^2)$$

and

$$uv_x + vv_y = \phi_x\phi_{xy} + \phi_y\phi_{yy} = \frac{1}{2}\frac{\partial}{\partial y}(\phi_x^2 + \phi_y^2).$$

so that

$$\rho(uu_x + vu_y) = \frac{\rho}{2}\frac{\partial}{\partial x}\|\mathbf{q}\|^2 = -p_x$$

and

$$\rho(uv_x + vv_y) = \frac{\rho}{2}\frac{\partial}{\partial y}\|\mathbf{q}\|^2 = -p_y.$$

Thus Bernoulli's law holds with the same constant everywhere in the case of irrotational, incompressible flow.

Subsonic flow of air around a wing can be modeled by irrotational, incompressible flow. Bernoulli's law is what provides the lift on a wing. The wing is shaped so that the air particles move faster over the top of the wing than across the bottom. Hence the pressure on top of the wing is less than on the bottom. The resulting pressure difference is the lift. The same phenomenon allows a sail boat to tack against the wind, a sail being a kind of vertical wing.

Many techniques have been developed to solve $\Delta\phi = 0$ to represent flow around an obstacle. If B is a body, which is impenetrable to the fluid, the velocity $\mathbf{q}$ must be tangent to the boundary curve of B. This means that on ∂B, $\mathbf{n}\cdot\mathbf{q} = 0$, where $\mathbf{n}$ is the exterior normal to ∂B. When the flow is irrotational and there exists a velocity potential ϕ, it must satisfy the boundary value problem

$$\Delta\phi = 0 \qquad \text{outside } B, \tag{12.33}$$

$$\frac{\partial\phi}{\partial n} = \mathbf{n}\cdot\nabla\phi = 0 \qquad \text{on } \partial B. \tag{12.34}$$

Example 12.6

The radial flow $\mathbf{q} = [x, y]/r^2$ is irrotational and incompressible with the velocity potential $\phi(x, y) = \ln r + C$. The streamlines are the radial lines $\theta = $ constant.

Example 12.7

We display the flow around an infinite cylinder whose cross section is the disk of radius a, $\{x^2 + y^2 \leq a^2\}$. We choose a velocity potential so that the flow far from the disk is almost uniform, $\mathbf{q} \approx [V, 0]$ where V is a constant:

$$\phi(x, y) = V\left(x + \frac{a^2 x}{x^2 + y^2}\right).$$

The stream function is

$$\psi(x, y) = V\left(y - \frac{a^2 y}{x^2 + y^2}\right).$$

On the boundary of the disk, $x^2 + y^2 = a^2$, so that $\psi = 0$. The velocity field is

$$\mathbf{q} = [\phi_x, \phi_y] = V\left[1 + \frac{a^2(y^2 - x^2)}{(x^2 + y^2)^2}, -\frac{2a^2 xy}{(x^2 + y^2)^2}\right].$$

The streamlines, which are the level curves of ψ, are plotted in Figure 12.6 using the commands contour. The shading of the figure represents the speed $\|\mathbf{q}\|^2$ according to the color bar at the right of the figure. Notice that $\mathbf{q} = 0$ at $(\pm a, 0)$. These are

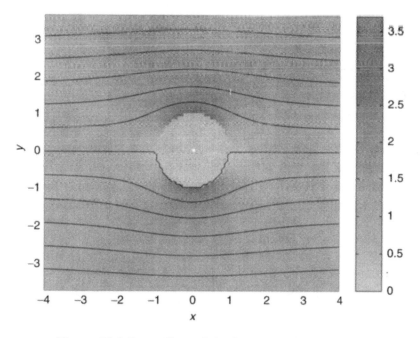

Figure 12.6 Streamlines of the flow around a cylinder.

called *stagnation points* in the flow. According to Bernoulli's law, the pressure is greatest at these points.

To make Figure 12.6, we shall take $a = V = 1$. We put the stream function and the speed squared in mfiles `stream.m` and `speed2.m`. Both of these functions become infinite at $(x, y) = (0, 0)$. Hence we cut them off, using the characteristic function of the exterior of the disk, (x.^2 + y.^2 > 1).

```
function z = stream(x,y)
    z = (x.^2 + y.^2 > 1).*(y - y./(x.^2 + y.^2));
   . . . . . . . . . . . . . . . .
function z = speed2(x,y)

    z1  = (1+ (y.^2 - x.^2)./(x.^2 +y.^2).^2).^2;
    z2 =  4.*x.^2.*y.^2./(x.^2 +y.^2).^4;
    z = (x.^2 + y.^2 > 1).*(z1 +z2);
   . . . . . . . . . . . . . . . . . .
```

```
>> x = linspace(-4,4,101);
>> y = linspace(-3,3,101);
>> [X,Y] = meshgrid(x,y);
>> pcolor(X,Y,speed2(X,Y));shading flat;
>> colormap(cool)
>> colorbar
>> levels = linspace(-3,3,11)
>> contour(X,Y, stream(X,Y), levels, 'k')
>> hold on
>> axis equal
```

We notice a jagged line where there should be a smooth curve for the level curve $\psi = 0$, which is the unit circle. To make a better picture, we change the set of level curves to be `levels = [linspace(-3, -.1, 6), linspace(.1,3, 6)]` and we plot the unit circle separately, inserting the instructions

```
>> t = linspace(0, 2*pi, 101);
>> plot(cos(t), sin(t))
```

before the `contour` command. The improved plot is shown in Figure 12.7. Remember, where the speed is greatest, the pressure is lowest.

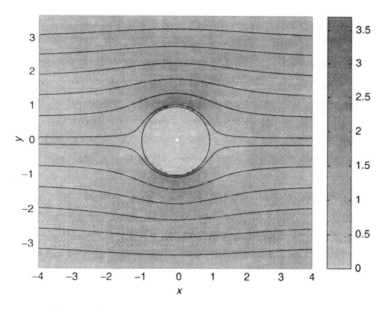

Figure 12.7 Improved picture of flow past cylinder.

Exercises

Problems in electrostatics

★ **1.** We use the method of images to produce the electrostatic potential $\phi(x, y, z)$ of a unit positive point charge located at $\mathbf{x}_0 = (0, 0, 1)$, with the plane $z = 0$ being grounded. That is, we want $\phi(x, y, 0) = 0$ for all x, y. To produce this potential, we use the fundamental potential $\gamma(\mathbf{x} - \mathbf{x}_0)$ at $\mathbf{x}_0$, balanced by another charge of the opposite sign at $\mathbf{x}_1 = (0, 0, -1)$. Thus we take

$$\phi(\mathbf{x}) = \gamma(\mathbf{x} - \mathbf{x}_0) - \gamma(\mathbf{x} - \mathbf{x}_1).$$

a) Verify that $\phi(x, y, 0) = 0$.

b) To get an idea of what the equipotential surfaces look like, plot the contours of $(x, z) \to \phi(x, 0, z)$ in the rectangle $R = \{|x| \le 5, \ 0 \le z \le 5\}$ using the command `contour`, with `levels = linspace(0,0.1,11)`. Now imagine these curves rotated around the z axis to form the equipotential surfaces. You can also use the mfile `impl` to view these surfaces. Describe in words what they look like.

c) Calculate the field $\mathbf{E} = -\Delta\phi$ and superimpose it on the plots of part b), using quiver and a much coarser grid. Multiply the vector field by the cutoff function that is zero on the disk $\{x^2 + (z-1)^2 \leq .5\}$. What is the direction of $\mathbf{E}$ on the plane $z = 0$?

★   **2.** Use the method of images to construct the potential $\phi(x, y, z)$ in the quarter-space $\{(x, y, z) : x, z \geq 0\}$ produced by a point charge located at $\mathbf{x}_0 = (1, 0, 1)$ that is zero on the planes $z = 0$ and $x = 0$. Put additional point charges at the symmetrically placed points $\mathbf{x}_1 = (1, 0, -1)$, $\mathbf{x}_2 = (-1, 0, 1)$, and $\mathbf{x}_3 = (-1, 0, -1)$.

a) Use the command contour to plot level curves of $\phi(x, 0, z)$ in the part of the quarter-plane $\{0 \leq x, z \leq 3\}$. Use levels = linspace (0,0.1,11).

Use impl to view the equipotential surfaces of ϕ. What do they look like?

b) Again compute the electric field $\mathbf{E} = -\Delta\phi$ and use quiver to superimpose it on the contour plot of part b). Use a much coarser mesh with quiver, and multiply the vector fields with the cutoff function that is zero on the disk $\{(x-1)^2 + (z-1)^2 \leq 1\}$. What is the direction of the field $\mathbf{E}$ on the surfaces $z = 0$ and $x = 0$?

★   **3.** A condenser is two parallel plates with opposite charges. In this exercise we shall make a numerical approximation to the potential of a condenser. The potential of a unit positive charge at $(x, y, 1)$ and a negative unit charge at $(x, y, -1)$ is

$$g(x, y, z) = \frac{1}{4\pi} \left[\frac{1}{\sqrt{x^2 + y^2 + (z-1)^2}} - \frac{1}{\sqrt{x^2 + y^2 + (z+1)^2}} \right].$$

Let R be the square $-1 \leq x, y \leq 1$. Now suppose the two plates are

$$P_{\text{upper}} = \{(x, y, z) : x, y \in R, \; z = 1\}$$

and

$$P_{\text{lower}} = \{(x, y, z) : x, y \in R, \; z = -1\}.$$

Suppose that both plates have a constant charge density σ. Then the potential we must estimate numerically is

$$\phi(x, y, z) = \sigma \int \int_R g(x - \xi, y - \eta, z) \, d\xi d\eta.$$

The potential can be approximated crudely by a Riemann sum. Choose points $\xi_j = -.75, -.25, .25, .75$ and $\eta_i = -.75, -.25, .25, .75$. Here $\Delta\xi = \Delta\eta = .25$. Then

$$\phi(x, y, z) \approx \sum_{i,j=1}^{4} \sigma g(x - \xi_j, y - \eta_i, z) \, \Delta\xi \, \Delta\eta.$$

a) Make a function mfile g.m for $g(x, y, z)$, taking $\sigma = 1$. Then write another function file, phi.m, that computes the Riemann sum for each x, y, z. Follow Example 12.1. You will need to make three-dimensional arrays [X,Y,Z] for plotting and two-dimensional 4×4 arrays [Xi,Eta] for the integration. Be careful in choosing your arrays for x, y, z so that x and y do not take on the values of ξ or η. For each i, j, k use the command sum(sum(g(X(i,j,k)-Xi, Y(i,j,k) -Eta, Z(i,j,k))))).

Check that your computed function $\phi = 0$ everywhere on the plane $z = 0$. This plane is an equipotential surface for ϕ.

b) Plot the function $(x, z) \rightarrow \phi(x, 0, y)$.

c) Combine a contour plot of $(x, z) \rightarrow \phi(x, 0, z)$ with a pcolor plot on the rectangle $\{-5 \le x, z \le 5\}$.

d) Calculate $\mathbf{E} = -\nabla\phi$ from the integral expression for ϕ. In what direction does $\mathbf{E}$ point on the plane $z = 0$?

4. Now imagine two plates, P_1 in the xy plane, $P_1 = \{-1 \le x, y \le 1, \ z - 0\}$, and one in the xz plane, $P_2 = \{-1 \le x \le 1, \ 1 \le z \le 3, \ y = 0\}$. Suppose that P_1 has a positive charge density σ and that P_2 has a negative charge density $-\sigma$. The potential is given by

$$\frac{\sigma}{4\pi} \int\int_{P_1} \frac{d\xi\, d\eta}{\sqrt{(x - \xi)^2 + (y - \eta)^2 + z^2}} - \frac{\sigma}{4\pi} \int\int_{P2} \frac{d\xi\, d\zeta}{\sqrt{(x - \xi)^2 + y^2 + (z - \zeta)^2}}.$$

Write a MATLAB code to make a Riemann sum approximation to the integral in the manner of Exercise 3. Plot the level curves of the potential in the xy plane, and in the xz plane.

5. Suppose a charge density $\rho(x, y, z) = x + y + z$ is carried on the cube $R = \{-1 \le x, y, z \le 1\}$. The resultant potential is

$$\phi(x, y, z) = \frac{1}{4\pi} \int\int\int_R \frac{\rho(\xi, \eta, \zeta) d\xi\, d\eta\, d\zeta}{\sqrt{(x - \xi)^2 + (y - \eta)^2 + (z - \zeta)^2}}.$$

Approximate the integral by a Riemann sum, and plot the level curves of ϕ in the xy plane, in the xz plane, and in the yz plane. What special symmetries do you see? What do the equipotential surfaces look like?

6. Suppose a charge is distributed uniformly on the z axis, $-\infty < z < \infty$ with a density λ. Let $\mathbf{E}$ be the electric field produced by this *line charge*. By a symmetry argument, we can see that the z component $E_3 = 0$ and that

$$\mathbf{E} = f(r)\left[\frac{x}{r}, \frac{y}{r}, 0\right],$$

where, as usual, $r = \sqrt{x^2 + y^2}$ and $f(r)$ is to be determined.

a) Apply Gauss's law to this field and a "pill box," which is a cylinder of radius a and thickness $2b$,

$$P = \{x^2 + y^2 \le a^2, \ -b \le z \le b\}.$$

There will be no contribution to the flux integral over the top and bottom of the pill box. Deduce that

$$f(r) = \frac{\lambda}{2\pi r}.$$

b) Show that the potential for this line charge is

$$\frac{\lambda}{2\pi}\ln(1/r)$$

so that the fundamental potential for a line charge is

$$\gamma_2(x, y) = \frac{1}{2\pi}\ln(1/r).$$

c) Verify that $\Delta\gamma_2(x, y) = 0$ for $r \neq 0$.

7. The fundamental potential of a line charge on the vertical line through (x_0, y_0) is $\gamma_2(x - x_0, y - y_0)$.

a) Use the method of images to construct the potential $\phi(x, y)$ for a line charge with $\lambda = 1$ located at $x_0 > 0$, $y_0 > 0$ and such that $\phi = 0$ on $x = 0$ and on $y = 0$.

b) Use a cutoff function and graph ϕ in the quarter-plane $x, y \ge 0$.

c) Use the `contour` command to plot the level curves of ϕ. What are the equipotential surfaces in three-dimensional space?

★   **8.** The *Green's function* for the disk $D_a = \{x^2 + y^2 < a^2\}$ is the potential ϕ from a unit (line) charge placed at a point $\mathbf{y}$ inside D_a such that $\phi = 0$ on the circle

$x^2 + y^2 = a^2$. This potential is also constructed by the method of images. Another unit charge with the opposite sign is placed at the point outside the disk:

$$\mathbf{y}^* = \frac{a^2 \mathbf{y}}{\|\mathbf{y}\|^2}.$$

a) Verify that $\|\mathbf{y}^*\| > a$ if $\|\mathbf{y}\| < a$.
b) The Green's function for the disk is

$$g(\mathbf{x}, \mathbf{y}) = \frac{1}{2\pi}\left[\ln\left(\frac{1}{\|\mathbf{x} - \mathbf{y}\|}\right) - \ln\left[\left(\frac{a}{\|\mathbf{y}\|}\right)\frac{1}{\|\mathbf{x} - \mathbf{y}^*\|}\right]\right].$$

Verify that $g(\mathbf{x}, \mathbf{y}) = 0$ for $\|\mathbf{y}\| < a$ and $\|\mathbf{x}\| = a$.
c) Set $a = 1$ and take $\mathbf{y} = (.5, 0)$. Use the contour command to plot the level curves of $g(\mathbf{x}, (.5, 0))$.
d) Compute the electric field $\mathbf{E}$ of this potential and superimpose it on the plot of part c). What is the direction of $\mathbf{E}$ on the boundary $x^2 + y^2 = 1$?

★   9. The *Poisson kernel* for the half plane $H = \{(x, y), y > 0\}$ is

$$P(x, y) = \frac{y}{\pi}\frac{1}{x^2 + y^2}.$$

a) Verify that $\Delta P = 0$ for $y \neq 0$.
b) Plot the level curves of the Poisson kernel in H.
The Dirichlet problem for H is

$$\Delta\phi = 0 \qquad \text{in } H$$

$$\phi(x, 0) = f(x)$$

for a given function $f(x)$ on the x axis. The solution of the Dirichlet problem in H is given by the integral

$$\phi(x, y) = \int_{-\infty}^{\infty} P(x - \xi, y) f(\xi)\, d\xi.$$

c) Write a script mfile that takes a function f defined in an mfile f.m, computes ϕ, and plots the surface. You can use the one-dimensional Simpson's rule to estimate the integral for each x, y. Assume that $f = 0$ for $|x| > 5$. Compute the values of the integral at the points in the meshgrid [X,Y] = meshgrid(x,y), where x = linspace(-5,5,51) and y = linspace(.05, 8, 51). We avoid the points $y = 0$ because the Poisson kernel is singular at the point $(0, 0)$, which will make $P(x - \xi, y)$ singular at $(\xi, 0)$. Let the variable of integration be xi =linspace(-5,5,101).

d) Let $f(x) = (x + .5) \exp(-.5x^2)$. You can cut off the integral to the interval $[-5, 5]$ because f is very small for $|x| > 5$. Graph ϕ and plot its level curves. Where are the maximum and minimum values of ϕ obtained?

e) Compare the computed values of the potential ϕ on the line $y = .05$ with the values of f. They should be nearly equal.

★   **10.** Solutions of the Laplace equation in the disk $\{x^2 + y^2 < a^2\}$.

a) Use the chain rule to verify that the Laplace operator $\Delta u = u_{xx} + u_{yy}$ in polar coordinates is given by

$$\Delta u = u_{rr} + \frac{1}{r}u_r + \frac{1}{r^2}u_{\theta\theta}.$$

b) Verify that $r^n \cos(n\theta)$ and $r^n \sin(n\theta)$ are solutions of $\Delta u = 0$.

c) Graph several of these solutions on the disk of radius $a = 1$, using polar coordinates (see Example 5.4).

d) Verify by hand that finite sums

$$u(r, \theta) = A_0 + \sum_{n=1}^{n=N}[A_n \cos(n\theta) + B_n \sin(n\theta)]r^n$$

are solutions of $\Delta u = 0$ with

$$u(1, \theta) = A_0 + \sum_{n=1}^{n=N}[A_n \cos(n\theta) + B_n \sin(n\theta)].$$

e) Let $f(\theta) = 2\cos(\theta) + \sin^2(\theta) + 4\sin(3\theta)$. Use a trig identity to expand $\sin^2(\theta)$ and find the solution of $\Delta u = 0$, $u = f$ on the circle $r = 1$. Graph the solution. Where are the maximum and minimum values on the disk $\{x^2 + y^2 \leq 1\}$ attained?

Problems in fluid flow

11. In this exercise we compare the Euler method of integrating the system (Eq. 12.14) with the more sophisticated method ode45 used in flow1.

Let the vector field be $q = [-y, x]$. The particle paths are circles. Write mfiles u.m and v.m for the components $u(x, y) = -y$, and $v(x, y) = x$. Write an mfile eulerflow.m for the script of Example 12.2.

a) Use the mfile eulerflow.m with starting point $(3, 0)$. Choose $\Delta t = .2$ and 30 time steps. The trajectory strays far from the circle of radius 3. Now use $\Delta t = .05$ and 120 time steps. Finally use $\Delta t = .01$ and 600 time steps. Compare the results.

b) Now use the mfile flow1.m, with corners being the vector $[-4, 4, -4, 4]$, and $T = 6$. Take $M = 1$ and click as closely as you can to the point $(3, 0)$. What is the computed path? How does this compare to the result for the Euler method? Note that for this integration, ode45 uses only 57 time steps. Although ode45 does a more complicated calculation at each step, the net result is that it is much more efficient than the Euler method.

12. Let the vector field be $q = [u, v]$, with $u(x, y) = x(1 - y)$ and $v(x, y) = 1$. Use flow1.m with corners for the square $0 \le x, y \le 4$. Use $T = 5$.

a) Plot trajectories starting from $(x_j, .5)$ with $x_j = 1, 2, 2.5, 3$. What kind of motion do you see?

b) Calculate the divergence of q. Where is $\text{div}(q) > 0$, and where is $\text{div}(q) < 0$? Do the particle paths give you much indication of the sign of the divergence?

13. Use the same vector field as in Exercise 12. Now use the mfile flow2.m with corners for the square $0 \le x, y \le 5$. Take times $t_1 = .3$, $t_2 = .6$, $t_3 = .9$, $t_4 = 1.2$ and click on the point $(2.75, .75)$. Try some other combinations of times and starting points. Observe the size of the images of the disk and their areas. Does this fit with the calculated value of $\text{div}(q)$?

14. Let the vector field $q = [u, v]$, where

$$u(x, y) = y, \qquad v(x, y) = -x + y(x^2 - 4)/5 + y^3/15.$$

a) Calculate the divergence of q. Where is it positive, and where is it negative?

b) Use the mfile flow1.m with corners for the square $-4 \le x, y \le 4$ and time $T = 10$. Use starting points on the x axis, $x = 1, 1.5, 2, 2.5, 3$, to get a feeling for the flow.

c) Now use the mfile flow2.m with the same corners. First choose times $t_1 = 1$, $t_2 = 2$, $t_3 = 3$, $t_4 = 4$, and click on the point $(1, 0)$. How do the images of the disk behave? Does this agree with the calculation of part a)?

d) Now use smaller times, $t_1 = .5$, $t_2 = 1$, $t_3 = 1.5$, $t_4 = 2$. This time click on $(2.5, 0)$, and watch how the disk deforms with the flow. Does this behavior agree with the calculation of part a)?

15. Return to Example 12.7, flow around a cylinder.

a) Find the points on the boundary of the cylinder, $x^2 + y^2 = 1$, where $\|\mathbf{q}\|$ is greatest.

b) Plot the contours of the velocity potential. Verify that they are normal to the circle and that ϕ is a solution of the boundary value problem (12.33), (12.34).

★   **16.** Consider the velocity potential $\phi(x, y) = x^2 - y^2$ and the stream function $\psi(x, y) = 2xy$. This pair of functions describes incompressible, irrotational flow in the quarter-plane, $x, y \geq 0$.

a) Verify that $\Delta\phi = \Delta\psi = 0$ and that $\nabla\phi \cdot \nabla\psi = 0$.

b) Plot the contours of ϕ and ψ in the square $\{0 \leq x, y \leq 2\}$. According to part a), what angle should their tangent vectors make where they cross?

c) Use the mfile `flow2.m` and times $t_1 = .2$, $t_2 = .4$, $t_3 = .6$, $t_4 = .8$. Put the center of the initial disk at $(x, y) = (.3, 1.4)$. Describe how the shape changes. What happens to the area? Is it conserved? Why?

d) Combine a `pcolor` plot of the speed squared, $\|\mathbf{q}\|^2$, with the contours of the stream function. Where is the stagnation point? According to Bernoulli's law, where is the pressure the greatest?

★   **17.** This exercise extends Exercise 16 to the case of a wedge of angle $\pi/3$. The velocity potential is $\phi(x, y) = x^3/3 - xy^2$, and the stream function is $\psi = x^2y - y^3/3$. Repeat parts a), b), c), and d) of Exercise 16. Use times $t_1 = .1$, $t_2 = .2$, $t_3 = .3$, $t_4 = .4$, and place the center of the initial disk at $(x, y) = (1.2, 1)$. When the program `flow2.m` is finished, use `hold on` and add the line $y = x\sqrt{3}$, $0 \leq x \leq 1$. This will indicate the edge of the wedge.

★   **18.** Let the velocity potential be $\phi(x, y) = \cos x \cosh y$ and the stream function be $\psi = -\sin x \sinh y$. Repeat the questions of Exercise 16 in the rectangle $\{0 \leq x \leq \pi, 0 \leq y \leq 1\}$. Use `flow2.m` with times $t_1 = .3$, $t_2 = .6$, $t_3 = .9$, $t_4 = 1.2$ and place the center of the initial disk on $(x, y) = (2.5, 1.5)$. Now there are two stagnation points. What kind of flow does this represent?

13

More Features of MATLAB

13.1 Data classes

There are four basic classes of data in MATLAB. We have been using all four of them without paying much attention. However, some very perplexing errors can occur if we are not careful. The classes are

Numeric   Symbolic   String   Inline

When we enter a number at the prompt ≫, this creates a numerical, double-precision quantity. For example,

```
>> x = 1/3;
>> y = 1/2;
>> x*y
ans =
     0.1667
```

Most of the examples we have considered involved calculation with numerical quantities.

As we have seen in Section 1.6, symbolic quantities are created using the command syms. For example,

```
>> syms x y
>> f = x*y + y^2
```

defines the symbolic expression $xy + y^2$. If we wish to make symbolic manipulations of constants as well as symbols, we must use the long syntax. For example,

```
>> x = sym('1/3')
>> y = sym('1/2')
>> x*y
ans =
      1/6
```

Compare with the numeric product x*y.

A string, often called a character array, is created when we enclose an alphanumeric sequence of characters in single quotes. For example,

```
>> z = 'a*x+5'
```

The string is a vector of numbers that represents the sequence of characters between the single quotes. You can look at the string components of z:

```
>> double(z)
ans =
      97    42   120    43    53
```

The numbers assigned to each character are the ASCII values.

Strings are often used to label functions or quantities. For example, in the definition of an inline function, we use strings:

```
>> f = inline('x.^2');
>> g = inline('x.^2 + exp(y)', 'x', 'y');
```

We can check the data class of each variable or quantity in the work space with the command whos.

Example 13.1

```
>> x = linspace(0,2*pi, 101);
>> y = x.^2;

>> syms a b c
>> d = 2*a*b+c^3;

>> w = '3/u +v';
>> f = inline(w,'u',v')

>> whos
```

Name	Size	Bytes	Class
a	1x1	126	sym object
b	1x1	126	sym object

c	1x1	126	sym object
d	1x1	142	sym object
f	1x1	876	inline object
w	1x6	12	char array
x	1x101	808	double array
y	1x101	808	double array

```
Grand total is 287 elements using 3024 bytes.
```

Converting from one class to another

There are several commands that convert data from one class to another. A symbolic expression, such as d in Example 13.1, may be converted into a string and used to define an inline function. The inline function may then be evaluated. Using d of Example 13.1, the conversion command is char(d). Then

```
>> g = inline(char(d), 'a', 'b', 'c');
```

is the inline function $g(a, b, c) = 2ab + c^2$. However, the inline function g may be used only on scalar values of a, b, c, and so cannot be graphed. To get a string expression that uses .* and .^, we use the command vectorize(d). Then

```
>> h = inline(vectorize(d), 'a', 'b', 'c');
```

is now array-smart and can take vectors and matrices for its f arguments. If we are interested only in graphing the symbolic function expression, we can use the command ezplot. We can do this without converting to an inline function.

Example 13.2

Let $f(x, y) = x \exp(-x^2 - y^2)$. We define f symbolically and compute Δf symbolically. We then convert the expression to an inline function for graphing and integration:

```
>> syms x y
>> f = x*exp(-x^2 -y^2);
>> laplacef = diff(f,x,2) + diff(f,y,2);
>> g = inline(vectorize(laplacef), 'x', 'y');
>> [X,Y] = meshgrid(-1:.05:2);
>> surf(X,Y,g(X,Y))
```

If you have the command ezsurf available, you can graph Δf without converting to an inline function.

To convert a symbolic expression to an mfile, first vectorize the symbolic expression, and then cut and paste into the mfile.

13.2   The command `feval`

The command `feval` is used to take the name of a function in the form of string, call the function, and evaluate it. For example, if $f(x, y) = 3x^2 - 2y$ is defined as an inline function, then

```
>> f = inline('3*x.^2 -2*y', 'x', 'y');
   >> feval(f,2,3)
   ans =
      6
```

Of course, we could have more easily written `f(2,3)`. If the same function f is given in an mfile `f.m`,

```
function z = f(x,y)
   z = 3*x.^2 - 2*y;
```

then we must use the name of f in the form of a string as the argument of `feval`.

```
>> feval('f', 2,3)
ans =
   6
```

The importance of `feval` comes in writing function mfiles that take functions as arguments. For example, the MATLAB rootfinder `fzero` has the call

```
fzero(fname, [x1,x2])
```

where `fname` is a string that is the name of a function. `feval` is used in the code of `fzero` to evaluate the function. Thus when f is given as an inline function, we use `fzero(f, [x1,x2])`, and when f is given in an mfile, we use `fzero('f', [x1,x2])`. This construction allows the function function `fzero` to accept any name for a function, f, g, h, F, G, etc. The same is true of the function mfiles, such as `simp2`, that are written for this text.

Example 13.3

You recall the function mfile `qsurf` that you can use to graph numerical functions of two variables over a rectangle $a \le x \le b$, $c \le y \le d$. The input arguments are the function name, and the 4-vector $[a, b, c, d]$.

```
function out = qsurf(f, corners)
   x = linspace(corners(1), corners(2), 51);
   y = linspace(corners(3), corners(4), 51);
   [X,Y] = meshgrid(x,y);
   Z = feval(f,X,Y);
   surf(X,Y,Z)
```

13.3 Vectorizing computations

Computations in MATLAB that involve loops or nested loops can often be speeded up if we take advantage of the many array operations of MATLAB. Our goal in MATLAB programming is to eliminate as many loops as possible.

Example 13.4

The operation sum is applied to a vector and adds up the components. Here are two codes to compute the sum $\sum_1^N n^p$. The long way is to do this in a loop:

```
function out = f(p,N)
    s = 0;
    for n = 1:N
        s = s + n^p;
    end
    out = s;
```

A more efficient way is to use the sum operator and the array operation .^:

```
function out = f(p,N)
    x = [1:N].^p;
    out = sum(x);
```

Example 13.5

The two-dimensional midpoint rule for integration over a rectangle R is

$$\int \int_R f \, dxdy \approx \sum_{j=1}^{n} \sum_{i=1}^{m} f(x_j, y_i) \Delta x \Delta y,$$

where (x_j, y_i) is the center of the subrectangle $R_{i,j}$ with sides Δx and Δy. The long way to implement this rule is to use nested loops:

```
function out = midpt(f,corners,n,m)
    a = corners(1); b = corners(2);
    c = corners(3); d = corners(4);
    delx = (b-a)/n; dely = (d-c)/m;
    x = linspace(a+.5*delx, b-.5*delx,n);
    y = linspace(c+.5*dely, d-.5*dely,m);
    s = 0;
    for i = 1:m
        for j = 1:n
            s = s + feval(f,x(j),y(i));
```

```
            end
        end
        out = s*delx*dely;
```

The double loop can be replaced with vector, matrix operations, which are more efficient. The operation sum, when applied to a matrix, sums down each column and puts the sum of each column in a row vector. Then applying sum a second time sums the elements in this row vector, yielding the sum over all the elements of the matrix. The last seven lines of previous code can be rewritten

```
        [X,Y] = meshgrid(x,y);
        F = feval(f,X,Y);
        out = sum(sum(F))*delx*dely;
```

F is the matrix of function values $f(x_j, y_i)$.

13.4   Programming

While loops

The programs we have written so far involved only simple for loops where the number of passes though the loop was determined in advance. In some circumstances, however, the number of passes may depend on the state of the computation. The notion of a *while loop* addresses this situation.

Example 13.6

Suppose our problem is to sum the terms of an infinite series, $\sum a_n$, on the computer. Of course, we sum only a finite number of terms and thereby make an approximation to the exact sum of the series. Our criterion will be that if $|a_n| < 10^{-6}$, we shall stop. Here is a short script that sums the series

$$e = \sum_0^\infty \frac{1}{n!} :$$

```
    term = 1;
    sum = 0;
    n = 1;
    while term > 10^(-6)
        sum = sum +term;
        term = term/n;
        n = n+1;
    end
```

Example 13.7

For a second example of a while loop, we return to Newton's method for numerically approximating the root of an equation $f(x) = 0$. This time, instead of specifying a certain number of iterations, we want to continue the process until a desired accuracy is achieved. We can estimate the error in Newton's method at a simple root by looking at the difference between two successive iterates, $x_{n+1} - x_n$. Here is a function mfile that implements Newton's method in one dimension. Its arguments are the functions f and f', the starting value, and the error tolerance. Compare this mfile with that of Section 7.2.

```
function out = Newton(f,fprime,xstart,tol)
   xold = xstart;
   xnew = xold - feval(f, xold)/feval(fprime,xold);
   while abs(xold -xnew) > tol
     xold = xnew;
     xnew = xold - feval(f,xold)/feval(fprime,old);
   end
   out = xnew;
```

Logic

Often a program must make a choice and proceed to make different calculations, depending on a parameter. The relevant commands are if, else, and elseif.

Example 13.8

Consider an income tax system with two brackets. For income less than $20,000 the rate is 10%, and for income in excess of $20,000 the rate is 15%. A short program to compute the tax could be written as follows:

```
function out = tax(income)
      if income < 20000
         out = .1*income;
      else
         out = 2000 + .15*(income - 20000);
      end
```

Notice that the end command is needed to close the if, else sequence. Now suppose the system has a third bracket, with a rate of 20% for income in excess of $50,000. We modify the function mfile tax as follows:

```
function out = tax(income)
      if income < 20000
```

```
          out = .1*income;
   elseif 20000 <= income < 50000
          out = 2000 + .15*(income-20000);
   else
          out = 6500 + .2*(income - 50000);
   end
```

One can add any number of branches with more `elseif` statements.

Another useful logical command is the `break` command, which terminates a loop when a certain value is reached in a computation before the index of the loop is exhausted.

Example 13.9

A Fibonacci sequence is a sequence a_n generated by the recursive scheme

$$a_{n+2} = a_{n+1} + a_n, \qquad n \geq 1.$$

To start the sequence, one must provide the starting values a_1 and a_2.

Input statements will be used to supply the starting values and the number of terms to be computed at run time. However, the first time $|a_n| > 1000$, the computation will stop. A script file that accomplishes this is

```
N = input('enter the number of terms to be computed  ')
a1 = input('enter a1  ')
a2 = input('enter a2  ')
% set aside a column vector of N memory spaces for the
% values of a.
a = zeros(N,1)
a(1) = a1; a(2) = a2;
for n = 1:N-2
    a(n+2) = a(n+1) + a(n);
    if abs(a(n+2)) > 1000
        break
    end
end
```

This program saves all the computed values in the column vector a. If you want to see a_{50}, enter a(50). If you want to see all of them at once, enter a.

For descriptions of other logical commands, enter `help or` or `help and`.

Appendix

Instructor Demos

In this appendix we discuss four mfiles that provide graphical displays that illuminate some important concepts. They are not intended to be numerical tools to make computations. They can be used by instructors in the classroom or by students on their own. Instructions on how to use the files can be found with the command `help`.

findroot.m minsurf.m flux3.m circ.m tarea.m

`findroot` is an interactive version of Newton's method to solve the system $f(x, y) = 0$, $g(x, y) = 0$. The program displays the zero-level curves of f and g in a rectangle where the root is sought. After clicking on the figure, the code plots intersections of the tangent plane approximations to f and g in the xy plane. Additional clicks on the figure solve the approximating linear system and plot the approximate solution.

minsurf uses the triangular patch representation of a surface to provide approximate solutions of Plateau's problem in a rectangle. minsurf uses the mfile tarea.m.

flux3 calculates the flux of a vector field through a square piece of surface. The piece of surface contains the origin in three-dimensional space, and it can be oriented in any direction.

circ calculates the circulation of a vector field around a circular disk. The disk is centered at the origin in three-dimensional space and can be oriented in any direction.

Solutions to Selected Exercises

Chapter 3

4. Place the center of the seat of the stool at the point $(0, 2.5)$. The points on the ground where the legs reach the circle of radius 1 can be chosen as $(1, 0, 0)$, $(-1/2, \sqrt{3}/2, 0)$ and $(-1/2, -\sqrt{3}/2, 0)$. The magnitude of the force on each leg is $(80/3)\sqrt{29}$.

5. The for loop when $n = 24$ is

```
deltheta = 2*pi/24;
t = [-1,1];
for theta = 0:deltheta:2*pi
    plot3(cos(theta)*t, sin(theta)*t, .7*t)
    hold on
end
```

11. Use a for loop. The spacing between the planes on the x axis is $\delta x = \cos(\pi/6)$. Let N be the normal to the planes.

```
for j = 1:8
    P = [(j-1)*delx,0,0];
    plane(P,N)
    hold on
end
```

13. The directions of the wires are $[-1/2, 0, -\sqrt{3}/2]$ for L_1, $[1/4, -\sqrt{3}/4, -\sqrt{3}/2]$ for L_2, and $[1/4, \sqrt{3}/4, -\sqrt{3}/2]$ for L_3.

Chapter 4

2. The polygonal approximation with 100 segments yields a value of 9.6869, while quad8 gives the more accurate value of 9.6885.

5. From the first component of $\mathbf{r}(t)$, find the time t_* when the projectile hits the ground, then substitute in the second component of $\mathbf{r}(t)$. This produces the formula for the range $R = v_0^2 \sin(2\theta)/32$.

9. The arc length of each rib is given by the integral $(1/4) \int_0^{8/a} \sqrt{1+u^2}\, du$, where $a = a(z)$. The integrals can be done by hand or symbolically and the values added up, or we can use quad8 on each integral and add up the results, as in this short script. This script yields a value of 202.6068.

```
z = 0:20;
a = -.0166*z.^2 + .2245*z + 2.25;
f = inline('sqrt(1+x.^2)')
length = 0;
for n = 1:21
        rib = (1/4)*a(n)^2*quad(f, 0, 8/a(n));
        length = length + rib;
end
```

11. Figure 4.11 shows the cam and cam follower when $\theta = \pi/3$. c) $l(\theta) = 1 + (1/2)\cos\theta$.

12. The minimum value of l is .4084, occurring at the values $\theta = 2.0106$ and $\theta = 2\pi - 2.0106$.

Chapter 5

4. h has the constant value $(1-a^2)/(1+a^2)$ on the line $y = ax$.

5. The error in the difference approximation to f_x is proportional to $|\Delta x|$.

7. d) $f_y(x, y) = x$ for $y > 0$ and $f_y(x, y) = -x$ for $x < 0$. $f_y(0, 0)$ exists and equals 0, because $f(0, y) = 0$ for all y.

8. a) The maximum value of $D_{\mathbf{u}} f(1, 1)$ is $\sqrt{5}$, attained in the direction $(1, -2)$. b) $D_{\mathbf{u}} f(1, 1) = 0$ when $\mathbf{u} = [2, 1]$.

9. The hottest spot is at the point $(.5, .5)$. The heat flux is always perpendicular to the level curves. The heat flows away from the hottest spot. On the left (insulated) edge, $u_x = 0$, which makes the flux vector parallel to the edge.

12. The error is reduced by factor of $1/4$ when h is halved.

Chapter 6

2. $c_* = 2/e$. The origin lies on S_{c_*}.

3. $101.2 \leq \rho(2, y, 0) \leq 101.7$ for $0 \leq y \leq 2$.

8. The arc length of the curve bounding the cross section of the wing is 4.1744, and the area of the surface of the wing is $4 \times 4.1744 = 16.6974$.

9. One set of mutually orthogonal vectors is $\mathbf{L} = [a, b, c]$, $\mathbf{u} = [ac, bc, -(a^2 + b^2)]$, $\mathbf{v} = [-b, a, 0]$. These work as long as $a^2 + b^2 > 0$. You should normalize $\mathbf{u}$ and $\mathbf{v}$. After you have created a meshgrid [S,T], the key commands are

```
X = L(1)*T + d*cos(S)*u(1) + d*sin(S)*v(1);
Y = L(2)*T + d*cos(S)*u(2) + d*sin(S)*v(2);
Z = L(3)*T + d*cos(S)*u(3) + d*sin(S)*v(3);
```

10. Let $\mathbf{p}(t) = (r \cos t, r \sin t, at)$ be the point on the helix, $0 \leq t \leq 4\pi$. Then the tube may be parameterized

$$\mathbf{x}(t) = \mathbf{p}(t) + (b \cos s)\mathbf{N}(t) + (b \sin s)\mathbf{B}(t),$$

where $0 \leq s \leq 2\pi$. The key MATLAB commands to graph the tube are similar to those in Exercise 9.

14. With the meshgrid [S,T] constructed, the key commands are

```
X = 2*cos(T) + S.*cos(T/2);
Y = 2*sin(T) + S.*sin(T/2);
Z = S.*sin(T/2);
```

A normal to the surface is given by $\mathbf{r}'(t) \times \mathbf{L}(t)$.

Chapter 7

1. There are six roots. The symbolic solver finds all of them as lengthy symbolic expressions. Rounding off to four digits, they are

$$(\pm 1.1169, \pm.8295), \quad (\pm.3836, \mp.9814), \quad (\pm.7780, \mp.9212).$$

3. Here are some of the numbers you should get, rounded to four digits. You can also check your numerical results by plotting the curve you compute together with the zero-level curve of f plotted by the contour command.

x	-1	$-.5$	$0.$	$.5$
y	.4797	.3323	$0.$	-1.1433

4. One root is easily spotted, $(0, 0)$. The other two are $(.3781, -.6149)$ and $(1.0858, 1.0420)$.

6. $A = 4.9510$, $B = .0980$, $C = .245 \times 10^{-4}$.

7. a) $\lambda_* = .7828$.   b) When $\lambda = 2$, there are two real roots, $(x_1, y_1) = (.8988, 2.6704)$ and $(x_2, y_2) = (1.2527, .8347)$.   c) When $\lambda = \lambda_*$, the single root is $(x_*, y_*) = (.8024, 1.3445)$.

8. b) If $|y|$ is large, the factor $(1 - 1/\sqrt{1 + y^2}) \approx 1$. Hence $c_1 \approx -(9.8/2)$. A good linear approximation is given by $y = -(9.8/2)m - 1$.

Chapter 8

2. g has a saddle point at $(x_0, y_0) \approx (-.33, .47)$.

5. f has saddle points at $(0, -1)$, $(1, 1)$, and $(-1, 1)$. It has minima at $(0, 1)$, $(1, -1)$, and $(-1, -1)$.

6.   c) Using `findcrit`, we see that f has a minimum at $(x_1, y_1) \approx (-1.191, -.180)$ and a maximum at $(x_2, y_2) \approx (1.322, .370)$. Using `newton2`, more accurate estimates are $(x_1, y_1) = (-1.19098, -0.17954)$ and $(x_2, y_2) = (1.32240, 0.30715)$.

8.   c) When the size of the square is halved, the error decreases (roughly) by a factor of $(1/2)^3 = 1/8$, which agrees with Eq. (8.7).

11. a) $\max_K f \approx 1.513$ occurs near $(\pm0.89, \pm0.33)$, and $\min_K f \approx -1.03$ occurs near $(\mp0.46, \pm0.63)$.   b) The symbolic solver finds the roots λ, x, y, and when evaluated in double precision and then rounded to 4 digits, the max occurs at $(\pm0.8881, \pm0.3251)$, with $\max_K f = 1.5490$. The min occurs at $(\mp0.4597, \pm.6280)$, with $\min_K f = -1.0490$.

14. The point $(x_0, 0, 0) = (75.9747, 0, 0)$ is a saddle point for $V(x, y, 0)$. A particle at a point $(x, 0, 0)$ with $x < x_0$ will fall toward the mass M_0. When $x > x_0$, the particle will fall toward the mass M_1. When $x = x_0$, the gravitational force $-\nabla V = 0$, so that the particle does not move.

15. a) When there is no regulation, the maximum profit is $.6172$, occurring at $(3.025, 2.0167)$. b) The rate of return is about $.24$ (24%). c) The set G_s lies to the right of the level curve $h = s$. d) The point of intersection moves down and to the right as s decreases. This means that capital is being substituted for labor as s decreases.

Chapter 9

1. The Riemann sum result with $m = n = 32$ yields 2.5762, and the analytic result is $-\sin(4) + 2\sin(2) = 2.5754$. The difference $\approx 8 \times 10^{-4}$.

3. a) Let $\mathbf{t}^x = (1, 2, 2, \ldots, 2, 1)$ be the $n+1$ vector of trapezoid coefficients in the x direction and $\mathbf{t}^y = (1, 2, \ldots, 2, 1)$ be the $m + 1$ vector of trapezoid coefficients in the y direction. The two-dimensional trapezoid method, applied to a function $f(x, y)$, is

$$T_{n,m}(f) = \frac{(b - a)(d - c)}{4mn} \sum_{i,j} t_j^x t_i^y f(x_j, y_i).$$

It can be rewritten

$$T_{n,m}(f) = \sum_{i,j} \bar{f}_{i,j} \operatorname{area}(R_{i,j}),$$

where $\bar{f}_{i,j}$ is the average of the function values at the four corners of the subrectangle $R_{i,j}$.

b)

$$\int \int_R g(x)h(y)\, dxdy = \int_a^b g(x)\, dx \int_c^d h(y)\, dy$$
$$= (T_n(g) + E_n(g))(T_m(h) + E_m(h))$$
$$= T_{n,m}(gh) + E_n(g)T_m(h) + E_m(h)T_n(g) + E_n(g)E_n(h)$$

When m and n are doubled, the error is reduced (roughly) by a factor of $1/4$, just as in the one-dimensional rule.

4. With $n = 40, m = 60$, Simpson's rule gives $I_1 = -0.38239636754005$. With $n = 80, m = 120$, Simpson's rule gives $I_2 = -0.38239213241589$. The exact result is $I = -0.38239185372996$. $|I_1 - I_2| = 4.235124158313841 \times 10^{-6}$ while $|I - I_1| = 4.513810089102651 \times 10^{-6}$. The difference $|I_1 - I_2|$ is a very good estimate of $|I - I_1|$.

7. With $m = n = 20$, the approximation is within .15 of the exact answer.

9. c) $a = -1.9954$ and $b = .6396$. d) The change of variable $x = u$, $y = v\sqrt{4 - u^2 - \exp(2u)}$ yields the integral $2 \int_a^b \int_0^1 q(u, v)\, dv\, du$, where

$$q(u, v) = (\sqrt{(1 - v^2)(4 - u^2)} + v^2 e^{2u} - e^u)\sqrt{4 - u^2 - e^{2u}}.$$

e) Simpson's rule with $m = n = 50$ yields $I_1 = 6.12134479405900$. Simpson's rule with $m = n = 100$ yields $I_2 = 6.12114816464481$. An estimate of the error in I_1 is $|I_1 - I_2| \approx 2 \times 10^{-4}$.

11. Making first the change of variable $\tilde{x} = x - 1$ and then using polar coordinates, $\tilde{x} = r \cos\theta, \; y = r \sin\theta$, we have

$$2 \iint_{(x-1)^2 + y^2 \le 1/4} \sqrt{4 - x^2 - y^2} \, dxdy = 2 \iint_{\tilde{x}^2 + y^2 \le 1/4} \sqrt{4 - (\tilde{x} + 1)^2 - y^2} \, dxdy$$

$$= 2 \int_0^{1/4} \int_0^{2\pi} \sqrt{3 - r^2 - 2r \cos\theta} \; r \, drd\theta.$$

Simpson's rule with $n = 10, m = 10$ yields $I_1 = 0.67601408747821$, while Simpson's rule with $n = 20, m = 20$ yields $I_2 = 0.67601407734176$, with a difference $I_1 - I_2 = 1.013645556380283 \times 10^{-8}$. The result I_1 is an estimate of the true value of the integral with an error on the order of 10^{-8}.

12. a) The mass of H is $2\pi b^2 \arctan(4)$. c) The mass of F is given by the integral

$$\iiint_F \frac{1}{1 + x^2} \, dV = 2 \iint_{x^2 + y^2 \le a^2} \frac{\sqrt{b^2 - y^2}}{1 + x^2} \, dxdy$$

$$= 2 \int_0^{2\pi} \int_0^a \frac{\sqrt{b^2 - r^2 \sin^2\theta}}{1 + r^2 \cos^2\theta} \; r \, drd\theta.$$

Simpson's rule with $m = n = 20$ yields $I_1 = 10.05448036681659$, while Simpson's rule with $m = n = 40$ yields $I_2 = 10.05433242388650$. The difference $I_1 - I_2 = 1.479429300896840 \times 10^{-4}$.

d) The mass of F is now

$$\iint_{\tilde{x}^2 + 2y^2 \le 2a^2} \int_{-\sqrt{b^2 - y^2}}^{\sqrt{b^2 - y^2}} \frac{dzd\tilde{x}dy}{1 + (z + \tilde{x})^2}$$

$$= \iint_{\tilde{x}^2 + 2y^2 \le 2a^2} [\arctan(\tilde{x} + \sqrt{b^2 - y^2}) - \arctan(\tilde{x} - \sqrt{b^2 - y^2})] \, d\tilde{x}dy.$$

Make the further change of variable $\tilde{x} = \sqrt{2}a \cos\theta, \; y = a \sin\theta$ so that the integral to compute becomes

$$\sqrt{2} \int_0^a \int_0^{2\pi} q(r, \theta) r \, drd\theta,$$

where

$$q = \arctan(a\sqrt{2}\cos\theta + \sqrt{b^2 - a^2\sin^2\theta}) - \arctan(a\sqrt{2}\cos\theta - \sqrt{b^2 - a^2\sin^2\theta})$$

Simpson's rule with $m = n = 20$ yields $I_1 = 8.67952960048467$, and with $m = n = 40$, $I_2 = 8.67952906940928$. The difference $I_1 - I_2 \approx 5 \times 10^{-7}$.

13. c) With $d = \sqrt{r^2 + a^2}$, the Jacobian determinant is $|J| = d + ru/d$ and the change of variable for x is $x(u, v, t) = (r + u)\cos t + (av/d)\sin t$. The integral for the mass is

$$\int_0^{8\pi} \int \int_{u^2 + v^2 \le b^2} (x(u, v, t) + 10)(d + ru/d) \, du\,dv\,dt.$$

Put in polar coordinates $u = \rho\cos\theta$, $v = \rho\sin\theta$. Using Simpson's rule with $n = m = 10$ and $p = 20$, we get $I_1 = 2652.380095534809$. With $n = m = 20$ and $p = 40$, we get $I_2 = 2652.410623115358$. The difference $I_1 - I_2 = .03052758054901$, which estimates the relative error as on the order of $.03/2652 \approx 10^{-5}$.

Chapter 10

3. b) Take $y(u, v) = uv$. The integral becomes

$$\int_0^1 \int_0^1 u\sqrt{1 + 4u^2(1 + v^2)e^{-2u^2(1 + v^2)}} \, du\,dv.$$

c) With $n = m = 10$, Simpson's rule yields $I_1 = 0.62217628062857$, and with $n = m = 20$, it yields $I_2 = 0.62218208324708$. The difference $I_1 - I_2$ is on the order of 6×10^{-6}.

5. a) The curved surface of the hull is the graph of $y(x, z) = (2x/a(z))^2 = 4x^2/a(z)^2$ over G. The surface area of this part of the hull is given by the integral

$$\int\int_G \sqrt{1 + y_x^2 + y_z^2} \, dxdz.$$

b) With the change of variable $x = a(z)u$, the integral becomes

$$\int_0^{20} \int_{-1}^1 \sqrt{a^2(z) + 64u^2(1 + u^2(a'(z)^2))} \, du\,dz.$$

c) With $n = m = 50$, Simpson's rule yields $I_1 = 195.6791$, and with $n = m = 100$ it yields $I_2 = 195.6795$, with $I_1 - I_2 \approx 3 \times 10^{-4}$. d) The area of the stern is

$$8a(0) - 2 \int_0^{a(0)} \left(\frac{2x}{a(0)}\right)^2 dx = 16a(0)/3 = 12.$$

7. b) The area is given by the integral

$$2\pi \int_{10}^{15} |f(v)| \sqrt{1 + (f'(v))^2},$$

where $f'(v) = -.36v^2 + 10.8v - 82$. c) quad8 gives a value of $1,275.8$ with a tolerance of 10^{-3}.

8. The hydrostatic pressure on the upper surface is given by the integral

$$\int\int_S p(z)\, dS = 62.5 \int\int_{x^2+y^2 \le h/4} (3h/4 + x^2 + y^2)\sqrt{1 + 4(x^2 + y^2)}\, dx dy$$

$$= 125\pi \int_0^{\sqrt{h}/2} r(3h/4 + r^2)\sqrt{1 + 4r^2}\, dr.$$

quad8 gives the answer 2.938×10^6 pounds, with a relative error of 10^{-3}.

10. The area integral is

$$\int\int_R \sqrt{1 + ((1 - 2x^2)^2 + 16x^2 y^2)e^{-2x^2 - 4y^2}}\, dx dy.$$

a) Simpson's rule works exceedingly well on this integral. With $m = n = 4$, $I_1 = 1.11519234133286$; with $n = m = 8$, $I_2 = 1.11509286114128$. The difference $I_1 - I_2 \approx 10^{-4}$.

b) Using a 5×5 mesh, tsurf gives a value of 1.06604860421834, which differs from I_1 by $\approx .05$.

12. Let A_j be the area of triangle j as labeled in the exercise:

$$A_1 = (1/2)\sqrt{17 + (4z - 6)^2}, \quad A_2 = (1/2)\sqrt{20 + (2z - 3)^3},$$

$$A_3 = \sqrt{(5 + (2z - 4)^2}, \quad A_4 = \sqrt{8 + (z - 2)^2}.$$

The minimim value of $A(z) = A_1 + A_2 + A_3 + A_4$ occurs at 1.7363. fmin finds this point with an error tolerance of 10^{-4}.

Chapter 11

1. b) With $n = 100$, Simpson's rule gives $I_1 = 42.0543$; with $n = 200$, Simpson's rule gives $I_2 = 42.0557$, with a difference $I_1 - I_2 \approx 10^{-3}$.

3. Using Simpson's rule on each segment with 50 subintervals, the result is -36.8082.

5. a) The line integral along these two segments yields a value of about 1.2 or 1.3 when done by eye. **c)** The path consisting of the two line segments $(.5, .5)$ to $(.5, 1.5)$ and then $(.5, 1.5)$ to $(2, 2)$ yields a value very close to zero, but nonnegative.

7. This vector field is obviously conservative.

14. b) The boundary integrals sum to 3.5. Simpson's rule is exact on polynomials of degree ≤ 3. **c)** Simpson's rule applied to the boundary integrals yields 3.8570168. simp2 applied to the double integral yields 3.8570169, with an error on the order of 2×10^{-7}.

Chapter 12

2. b) On the plane $x = 0$, the electric field **E** points in the direction $(-1, 0, 0)$; on the plane $z = 0$, **E** points in the dirction $(0, 0, -1)$. In both cases, it points to the exterior, away from the positive charge.

3. e) We can calculate $\nabla\phi$ by differentiating under the integral as long as (x, y, z) does not lie on either plate. Now, $g_x((x-\xi), (y-\eta), 0) = g_y((x-\xi), (y-\eta), 0) = 0$. Hence $\phi_x(x, y, 0) = \phi_y(x, y, 0) = 0$. Thus

$$\mathbf{E}(x, y, 0) = (0, 0, -\phi_z(x, y, 0)) = -\frac{1}{2\pi} \int\int_R \frac{d\xi\, d\eta}{[(x - \xi)^2 + (y - \eta)^2 + 1]^{3/2}}.$$

The electric field points down from the positively charged plate to the negatively charged plate.

6. a) The flux through the cylindrical boundary S is

$$\int\int_S \mathbf{E} \cdot \mathbf{n} \, dS = f(a) \int\int_S dS = 4\pi abf(a).$$

By Gauss's law, this flux must equal the charge contained in the pill box, $2 > b$.

Equating these expressions, we get $f(a) = \lambda/(2a\pi)$. Since a is arbitrary and the answer is independent of b, we deduce that $f(r) = \lambda/(2\pi r)$.

9. Simpson's rule does not do a good job for y close to zero because the integrand is singular. However, it gives some idea of the solution. f and P should be defined in mfiles or as inline functions. Then a script file to compute ϕ and plot it is given by

```
n = 50; m = 100;
x = linspace(-5,5,n+1); y = linspace(.05, 8,n+1);
xi = linspace(-5,5, m+1);

s = simpvec(m); h = 10/m;

[X,Y] = meshgrid(x,y); phi = zeros(size(X));

for j = 1:m+1
  phi = phi+(h/3)*s(j)*f(xi(j))*P(X-xi(j),Y);
end

surf(X,Y,phi)
pause

plot(x, f(x), x, phi(1,:), 'r')
pause

levels = linspace(-.4, 1, 41);
contour(X,Y,phi, levels)
```

The maximum and minimum values of ϕ are attained on the line $y = 0$ and are the same as the maximum and minimum values of f.

10. d) The solution is

$$u(r, \theta) = \frac{1}{2} + 2r\cos(\theta) - (1/2)r^2\cos(2\theta) + 4r^3\sin(3\theta).$$

The maximum and minimum over any disk $\{x^2 + y^2 \le a^2\}$ are attained on the boundary circle $\{x^2 + y^2 = a^2\}$.

15. $\|\mathbf{q}\|$ is greatest at the points $(0, \pm 1)$.

16. c) Although the shape changes, the area remains constant because the flow is incompressible. d) The origin is the stagnation point and the pressure is greatest there.

Index

291

LaVergne, TN USA
29 January 2011
214493LV00004B/11/P

9 780121 876258